NATIONAL SECURITY
SPACE STRATEGY CONSIDERATIONS

Rick Larned, BGen., USAF Ret.
Cathy Swan, Ph.D., Col., USAF Ret.
Peter Swan, Ph.D., LtCol., USAF Ret.

National Security
Space Strategy Considerations

Published by Lulu.com
dr-swan@cox.net

ISBN 978-0-557-31774-5
Printed in the United States of America

TABLE OF CONTENTS

PREFACE

We are pleased to be able to introduce this monograph to you. The subject of National Security Space Strategy has long been of interest to us. In the time that we have been associated with space policy, requirements and programs, it has become apparent that our interest in Space Strategy has lagged. President Eisenhower called for "freedom of space" and "space for peaceful purposes" 50 years ago. That is still an objective worth shooting for. Our achievements in space since then easily fit in the top ten marvels of the 20th century. The commitment involved in those accomplishments reflected the priority and needs of our nation at the time. In those days, our acquisition strategy was simple: support demanding national requirements by making large leaps in technology. Our operational strategy was also simple: accomplish great things by sheer perseverance and determination to overcome each hurdle.

Today, space has matured into an operational medium in its own right - one that users throughout the World depend upon. Our goals remain those of President Eisenhower. The challenges to achieving those goals are growing daily. Developing a National Security Space Strategy to accomplish those goals is more important than ever before.

This monograph brings together a collection of strategy

considerations that should be helpful to all sides of the debate. Many of the discussions bring out points that lead toward a balance in our national strategies. We hope it provokes lively discussion.

ACKNOWLEDGMENT

Several people made significant contributions to this monograph. Colonel Gary Dahlen (USAF, Ret.) provided insightful comments based upon his extensive experience in the space operations world, in U.S. Space Command, in the National Reconnaissance Office (NRO), and in the Aerospace Corporation. Several space experts as well as principals in Burdeshaw Associates, Ltd., contributed valuable advice for which we are extremely grateful. Thanks, too, to Mr. Jesse Jarvis and Ms. Kathy McClurg, of Burdeshaw Associates, Ltd., for their exceptional graphics talent.

The authors, of course, are solely responsible for any mistakes in this monograph. And if you take exception to any of the points made, we'd be delighted to discuss them with you.

To the stars and beyond...

Chapter 1

WANTED: NATIONAL SECURITY SPACE STRATEGY

INTRODUCTION

America Needs a New National Security Space Strategy. The Nation is struggling with the future of National Security Space (NSS), and is at a critical juncture in being able to divine, budget and implement a strategy that will balance the essential elements of space dominance. The last comprehensive study of space strategy, "Space System Architecture 2000" (classified SECRET), focused on the Air Force, with other Service and Agency implications added where appropriate. There have been numerous studies and reviews of U.S. National Security Space programs since then that have reinforced the call for a new strategy. The most recent of these, a review of National Space Strategy, was conducted by an independent panel, signed by 30 space notables, and forwarded to President Obama.

> "We as a group of concerned citizens and space professionals believe the nation's space enterprise is a critical strategic foundation of our country's economic

well-being and technological vitality, as well as our overall national power and international standing. While the United States still leads the world in space, there are numerous problems that threaten our continued leadership. We face near-term mission gaps in important space capabilities; the U.S. space industry and workforce is losing its competitive edge; our engagement and influence in international space activities has declined; and there is widespread program overreach – that is, recurring cost overruns and delays, and more government space programs than the federal budget can currently support. In many respects, the nation's civil, commercial, defense, and intelligence space sectors are in worse condition today than they were a decade or more ago."[1]

In 2008, the Young Commission called for the development and execution of a national space strategy, as well as the reestablishment of "the National Space Council chaired by the National Security Advisor to implement the strategy and coordinate activities for NSS across the DoD, Intelligence Community, NOAA, NASA, and other responsible agencies."[2] This monograph summarizes issues that should be considered in the development of the segment of national space strategy that deals with National Security Space programs.

It is Time for an Upgrade. Independent of the specific focus of each study, and regardless of a study's sponsor, a consistent finding has been the need for a broader agreement

[1] "America's Leadership in Space," letter to President Obama from the Committee for U.S. Space Leadership, March 10, 2009

[2] Mr. A. Thomas Young, Chairman, "Leadership, Management, and Organization for National Security Space (NSS) – Report to Congress of the Independent Assessment

on a clear, comprehensive NSS strategy. However, while each study has called for its development, none has made even an illustrative attempt to describe such a strategy.

> "Despite operating in space for 51 years, the United States still lacks a comprehensive space strategy."[3]

BACKGROUND

President Eisenhower set the stage. Some have even gone so far as to say the U.S. does not have a National Security Space Strategy. In fact, while it is probably fair to say that we have not had a formally enunciated strategy, we have, in fact, been following a space strategy unerringly for 40 years. This strategy is manifested in the way space forces have been going about their business over that time. It is larger than a single DoD military space strategy or a single Intelligence Community space strategy; it includes both, plus the cross-cutting elements of civil and commercial space. President Eisenhower expressed his expectation of National Security Space Strategy with two tenets: "Freedom of Space," and "Space for Peaceful Purposes."

The National Security Space initiatives of President Eisenhower's Administration covered a breathtaking array of space capabilities. The Nation's communications, navigation, missile warning, and meteorological programs were all getting up to speed in the "White World," as were reconnaissance programs in the National Reconnaissance Office. In those days

---

Panel on the Organization and Management of National Security Space," Institute for Defense Analyses, IDA Group Report GR-69, July, 2008.

[3] AU-18 Space Primer, ACSC, Air University Press, Maxwell AFB, Alabama, 9/09. pg 39.

reconnaissance had a high priority within the Intelligence Community. Performance was key while cost and schedule took a back seat. As with other programs whose development was deemed key to national security, funding was "whatever it took" to be successful. Program sponsorship was at the highest level in the Executive Branch, and congressional oversight was typically limited to two members per committee. Detailed oversight was virtually non-existent. At the same time, the utility of Intelligence, Surveillance, Reconnaissance (ISR) products, such as imagery, was usually limited to Intelligence Community (IC) users who could wait for the images to be dumped from a satellite, caught in mid-air by an airplane, flown to Washington DC, interpreted by IC analysts, and finally delivered by car to the President.

Real-time systems changed the landscape. As overhead sensors became more sophisticated, the utility of space ISR grew commensurately. The advent of real-time ISR systems introduced an aspect of space products that was of great interest to warfighters. The systems providing such products, though, were not designed for real-time utility, and suddenly the community had to deal with military as well as IC users. Similarly, other "force enhancement" space services – communications, weather, navigation, missile warning – steadily provided greater and greater utility to more users.

Technology sets the pace. Because of the importance of National Security Space programs to the well-being of the United States, our Government has been willing to take tremendous technological risks over the last 50 years. This willingness to recognize the inextricable link between risk and reward has paid off in state-of-the-art achievements. In the early days, each rocket launch was a fingers-crossed adventure. Each communication to a new satellite was miraculous. "Staying the course" in the Corona film reconnaissance program through 12 successive failures led to a capability that was remarkable then and eye-opening even today. Each successful space program raised the bar that much higher, challenging everyone to push risk areas and leverage funding and schedule to do even better.

The users of NSS programs benefited from this determination to push the envelope. Largely because of technology limitations, early space programs were hard-pressed to do more than support a few strategic users. As our capabilities improved, tactical users became aware of NSS, first as a novelty, then a utility, and now as an essential part of their combat operations. Today, NSS programs support strategic and tactical users almost continuously with weather, navigation, communications, missile warning and intelligence products such as mapping, near-real-time signals intelligence (SIGINT), and processed imagery.

Nevertheless, in almost all cases, the U.S. Government (USG) followed a strategy for National Security Space programs that relied on (1) budgets large enough to solve impossible problems, (2) a close working relationship with industry to develop programs that would meet extraordinary technical challenges, and (3) satellites designed "better than spec" that could last longer than expected, and which provided a cushion while policy-makers debated the next generation of replacement programs.

PROBLEM

Yesterday's strategy is insufficient for today's space forces. These successes evolved from a robust technology environment supported by funding profiles sufficient to get the job done. Over time, they led to the National Security Space Strategy that we are following today:

- Exploit the advantages of the High Ground of space to the maximum extent possible,
- With the best systems possible,
- In as timely a manner as possible,
- Under increasingly stringent cost guidelines,
- And all the while improving support to DoD and Intelligence Community users, as well as to our friends and allies.

Today's National Security Space Strategy looks backward, not forward. Essentially, this strategy relied upon three critical aspects of the development and operation of space programs.

- **The Acquisition Phase was focused on performance.** Recognizing that space really is "rocket science," the USG allocated sufficient budgets over many years to ensure that industry could provide the systems to support strategic and tactical forces. This ultimately led to space programs that gave our forces capabilities unmatched by any other nation in the world.
- **The Operations Phase was Manpower Heavy.** The development of these unique, technology-rich systems required large teams of people on Earth to operate the satellites as well as leverage the massive amounts of data flowing from them. This workforce-rich environment usually equated to two teams at each ground station:

one to keep the satellites operating and another to ensure efficient flow of mission data. The data then went to analysts who made sure that the users in the field or in analysis centers received it.

- **The Sustainment Phase was Managed by Default.** Sustainment is an essential element of all weapon systems. Whether forces are on the ground, at sea, or in the air, personnel and logistics are constant challenges. People need to be recruited, hired, trained and mentored. Facilities and hardware on Earth need to be maintained, protected, upgraded, and monitored.

What is different with space programs, though, is that for all but a few isolated cases, the satellites themselves are outside the reach of depot maintainers, logisticians or unit crew chiefs. For NSS programs, "sustainment" of a satellite is essentially limited to how clever the software engineers are at re-programming satellite operations when something goes wrong. Beyond that, satellite "sustainment" means living with whatever lifetime you can get out of the hardware.

Fortunately, the NSS community has been graced with an unexpected bonus from America's satellite programs, one that perennially vexes budgeteers who are trying to calculate when to start funding replacement programs. This characteristic of space operations has defied our ability to quantify or predict lifetime of the satellites. It turns out that once a satellite has survived the initial hazards of getting into orbit, it tends to live much longer than its calculated lifetime. This extended-life benefit has been critical over the years in order to cover potential gaps when replacement programs are delayed in development.

A good strategy should not be based on excess and luck. In today's economic environment, it is unrealistic to count on sufficient funding to anticipate challenging developmental problems, a surfeit of manpower, and unexpectedly extended lifetimes for programs whose performance is critical to national security. The issue is not that America lacks a strategy for National Security Space programs; rather, the issue is how to change that strategy in order to accommodate economic and military realities in the future. There are legitimate obstacles that keep credible groups of space experts from actually defining how to bring our NSS strategy into the 21st Century. Suffice it to say, however, that solutions to these obstacles must be clearly and decisively recognized and understood if our space forces are to Protect and Serve our nation, our allies, and our friends.

One reason we haven't done well in updating our strategy may be that the employment of space programs is fundamentally different from the employment of other National Security forces. With "traditional" forces, a weapon system is declared "operational" after it has been designed, built, tested and deployed. After completing its years of productive service, the weapon system is typically decommissioned, scrapped and sold for parts, stored for future possibilities, and/or sold to our allies.

Satellites rarely follow this pattern. Satellites frequently support National Security objectives well before full operational capability (FOC) and even initial operational capability (IOC). At the other end of their duty, it is sometimes difficult to tell when a satellite is actually "done." Instead, residual capabilities are exploited for as long as possible. In some cases, that additional lifetime has carried well beyond the time when a satellite had run out of fuel for maneuverability.

A good Strategy should encompass all users. The market for space applications is growing exponentially. Today, users include a much wider swath of military warfighters and

Intelligence Community analysts. "Virtually all aspects of military operations are affected in some way by the capabilities provided from (space and cyberspace), and it's difficult to overstate their importance to the success of our Armed Forces," said Gen. Norton Schwartz, the Air Force Chief of Staff.[4] Users throughout the USG, industry, and society have an important stake in the future of National Security Space as well.

A good Strategy needs to address three parts of a satellite's lifetime. An Acquisition Strategy should cover how to buy and build space programs that will provide as much capability as can be squeezed into a given platform. An Operational Strategy should cover how to maximize support to users, as well as how to protect and defend our space forces. A Sustainment Strategy should cover how to strengthen our space forces for the long term. The National Security Space Strategy considerations proposed in this monograph address all three.

A good Strategy should drive investment decisions. In the sense of an implementing document for budget apportionment, Policy doesn't go far enough. Strategy needs to link Policy to Budget.

A good Strategy must be based upon the Big Picture. Thus, comprehensive National Security Space Strategy is based upon the following basic "OV-1" Operational View. This diagram shows how NSS Strategy fits inside the USG planning process that starts with policy at the highest level and goes all the way down to implementation recommendations within each organization.

---

[4] "Chief of staff highlights importance of space to Air Force mission," Defense Media Activity-San Antonio, February 19, 2010.

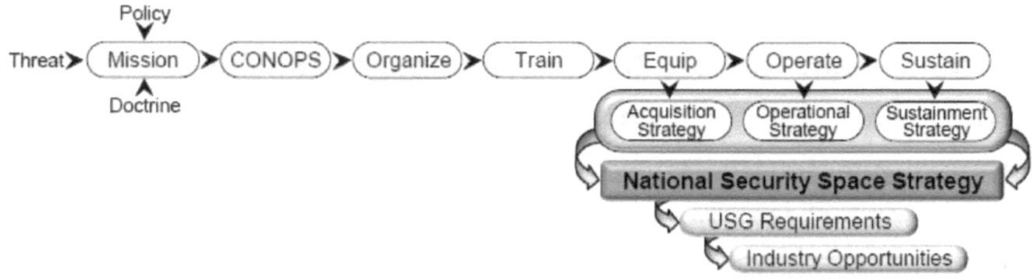

There are many ways to make this picture more accurate or more detailed. At the most basic level, the conjunction of Policy, Doctrine and Threat determines Missions for all national security forces, space included. A Concept of Operations (CONOPS) is then developed that shows how to accomplish each Mission. Once the forces supporting that CONOPS Organize, Train and Equip their units, the units then deploy and execute the Mission.

As shown in the OV-1 figure, this approach suggests that National Security Space Strategy should include three overarching components: Acquisition Strategy, Operational Strategy, and Sustainment Strategy. Each of these three components is essential to a complete NSS Strategy and will be examined in greater detail. Moreover, we have included Measures of Effectiveness (MOE) for each of these components. We also identify USG "Needs" that flow from these components, as well as Industry Opportunities that flow from these "Needs."

NATIONAL SECURITY SPACE

"National Security Space" means different things to different people. For purposes of this monograph, NSS is generally considered to have five components:

- **Department of Defense** (DoD) space programs include missile warning satellites (DSP, SBIRS), navigation satellites (GPS), communication satellites (many), and weather ("meteorological") satellites (DMSP).

- **Intelligence Community** (IC) satellites were euphemistically called "National Technical Means of Verification" in the past. Today most IC programs fall into the "ISR" (Intelligence, Surveillance, Reconnaissance) category and are classified, with products usually shared with the DoD.

- **Civil** space programs that support National Security missions are typically meteorological satellites such as GOES, NPOESS and others. In addition, NASA and NOAA lead efforts around the world to collect science data with Earth-based satellite systems as well as across the solar system to discover new truths about global warming, Earth's basic functions, galactic origins, and basic physics, chemistry, and biology. In addition, NASA runs the human exploration program with its International Space Station and missions to Mars.

- **Commercial** space programs used by USG organizations primarily include communication satellites. In addition, the USG participates heavily in a growing market in commercial imagery. The United States has always led private investment in commercial space systems. Historically, commercial communications satellite systems have been very successful and started competition all around the world, both in Geosynchronous Earth Orbit (GEO) and in Low Earth Orbit (LEO). Today commercial imagery satellite systems are developing business models and trying to capture major portions of the space marketplace. In the future, the human spaceflight equation will be altered with commercial suppliers – first to space in suborbital flights, and then to the International Space Station as a shuttle service.

- **Launch Infrastructure** is responsible for getting everything into orbit. Included are the rockets themselves, their upper stages (if used), launch complexes, and range tracking systems. Launch is exceedingly important in a strategic sense. Launch facilities undergo extensive stress due to exposure to the toxic environment of each launch, as well as the harsh effects of coastal weather damage at the launch facility itself and downrange. Natural as well as hostile vulnerabilities abound and need constant attention. Launch infrastructure includes three separate communities: DoD/Intelligence, civil (NASA), and commercial launchers and their associated support infrastructures.

The Core of NSS, called "Big Space," includes the major DoD and Intelligence Community satellite programs. Together with their commercial and civil complements, NSS is providing global leadership in key services from space, while maintaining a robust supplier base for the Nation.

- **Missile Warning.** The DoD Defense Support Program has monitored missile launches around the world for years. Its replacement, SBIRS, should improve the quality and timeliness of reporting by an order of magnitude.

- **Navigation.** Precision navigation from GPS is so automatic today that the Warfighter uses it without even thinking about its space origins.

- **Communication.** Likewise, communications move by satellite at the speed of light around the globe, 24/7. When there is an occasional communications outage, backups are in place to keep interruptions to a minimum. Today a soccer mom using a cell phone probably suffers more dropouts while driving the back roads of Virginia than a Warfighter experiences in the interior of Afghanistan. Naturally, there will always be a demand for more communications. However, satellite communications and their terrestrial counterparts provide better connectivity worldwide than any other military force has ever had – and possibly, if you listen to some of the pundits, too much.

- **Weather.** In the same vein, weather support provided by military, civil and commercial systems is available whenever and wherever the Warfighter needs it. Just as motorcyclists can rely on www.weather.com to plan their daily commute, military and commercial pilots don't take off without first checking satellite weather. There are several historical examples of the use of civil meteorological satellites (metsats) to back up DMSP when military weather satellite data were not available. This cross-stovepipe application of utility is highly desirable and should be copied for other space support elements.

- **ISR.** The Intelligence, Surveillance, Reconnaissance support provided by satellites is more complex. Satellites have provided a variety of kinds of ISR support over the years. With the growing capabilities of today's systems, roles and missions in the ISR world are beginning to overlap. At an admittedly basic level, much of today's ISR can be categorized as intelligence collection (such as signals intelligence (SIGINT) and imagery intelligence (IMINT)), surveillance (such as DSP's missile warning discussed above) and reconnaissance (provided by a variety of space-based and non-space systems, both DoD-sponsored and IC-sponsored). From an orbitology standpoint, satellites in lower orbits tend to provide more fleeting coverage, but with greater resolution, because they are closer; those in higher orbits provide more sustained coverage, but with less resolution. A comparison of SIGINT and IMINT satellites might make their operational differences clearer.

- **SIGINT.** Signals intelligence satellites use many different orbits and have many different operating parameters. From an operational standpoint, they can only cover the signals they are designed to cover, and then only when they are in position to collect such signals. Typically, though, their tasking-collecting-processing-disseminating cycle is very short. While improvements in geolocation and other operating parameters are always possible, Warfighters tend to be more satisfied with the persistence and responsiveness of SIGINT systems than with IMINT systems.

- **IMINT.** In order to get the resolution needed, IMINT satellites must fly at lower operating altitudes, which means the satellites can see a target on the ground for only a few minutes as they pass overhead. Because an imaging satellite has to be pointed at a specific target in order to "see" it, IMINT systems also have to put up with a "sophisticated" (read "occasionally cumbersome") targeting process. It is not uncommon to have conflicting target requests, which means that IMINT systems may not be able to provide the responsiveness typically needed by a military unit in combat. They need help in seconds rather than in minutes, hours or days. The truth is that for the persistence and responsiveness required by a force engaged in combat, space is usually not the best place to look. To provide "in seconds" responsiveness to changing battle conditions on the ground, a satellite must already be providing the necessary support, and in real time. Changing a satellite's tasking normally takes too long to meet a "troops engaged" scenario. If coverage isn't already part of the battlefield commander's arsenal, it is usually too late to provide additional help from space. For tactical responsiveness in a firefight, terrestrial forces (UAVs, aircraft and helicopters in the area, even troops on the ground) can probably provide better "I need it right now" responsiveness than satellites can.

NEXT

You should now have at least a cursory understanding of what "National Security Space" encompasses, and why a new strategy for NSS programs is desirable. With that as background, the next step is to show how NSS Strategy fits into the "Big Picture."

Chapter 2

HOPEFUL POLICY, UNFINISHED DOCTRINE

INTRODUCTION

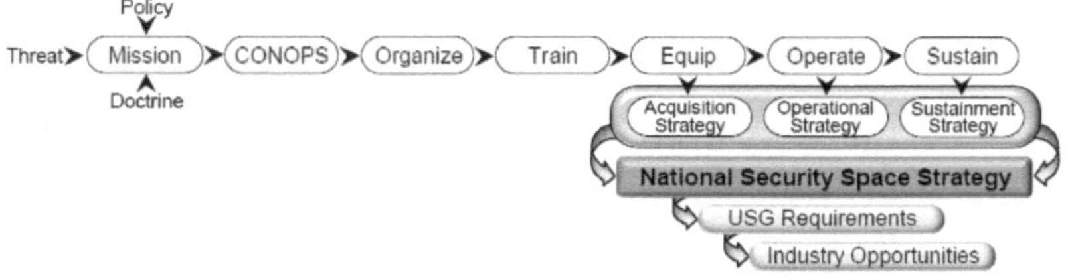

Policy and Doctrine set the stage for Strategy. National Security Space Strategy is developed within a framework of national and international considerations. Two of those considerations, National Space Policy and Military Space Doctrine, form binding constraints within which missions, strategies and operational concepts are formulated.

HOPEFUL POLICY

U.S. policy has always depended upon U.S. industry participation in all five segments of space activities (DoD,

Intelligence, civil and commercial communities, and launch infrastructure). This support has led to U.S. leadership (with a significant portion of the total space market) in the global space economic environment, currently valued at $251 billion per year.[5] This chapter addresses weaknesses in today's Policy and Doctrine strictures.

National Space Policy, 2006, was a good step forward. In 2006, the President authorized a new national space policy that asserted, "Those who effectively utilize space will enjoy added prosperity and security and will hold a substantial advantage over those who do not. Freedom of action in space is as important to the United States as air power and sea power. In order to increase knowledge, discovery, economic prosperity, and to enhance the national security, the United States must have robust, effective, and efficient space capabilities."[6] Commensurate with this declaration, the National Space Policy set the following goals:

- Strengthen the nation's space leadership and ensure that space capabilities are available in time to further U.S. national security, homeland security, and foreign policy objectives.
- Enable unhindered U.S. operations in and through space to defend our interests.
- Implement and sustain an innovative human and robotic exploration program with the objective of extending human presence across the solar system.

---

[5] Global Space Activity Revenues and Budgets, 2007. The Space Report. www.thespacereport.org.

[6] U.S. National Space Policy, August 31, 2006, p.1

- Increase the benefits of civil exploration, scientific discovery, and environmental activities.
- Enable a dynamic, globally competitive domestic commercial space sector in order to promote innovation, strengthen U.S. leadership, and protect national, homeland, and economic security.
- Enable a robust science and technology base supporting national security, homeland security and civil space activities.
- Encourage international cooperation with foreign nations and / or consortia on space activities that are of mutual benefit and that further the peaceful exploration and use of space, as well as to advance national security, homeland security, and foreign policy objectives."[7]

The 2006 Space Policy left holes that need to be filled. Policy, an essential building block in the development of Strategy, establishes guidelines within which National Security Space operations can be conducted. Rarely rigid, Policy is flexible and can change with a new leadership, Administration, or other political decision-making process. One of the more reassuring aspects of National Space Policy is that while policy has changed over the years as it has captured special issues, over-arching space policy itself has been remarkably consistent. Currently, the Obama Administration is preparing new policy, and a Space Policy Review is underway. Strategy must therefore remain closely attuned as the political winds of Policy shift.

---

[7] U.S. National Space Policy, August 31, 2006, p.2

What has happened is the growing recognition that people's concerns about an alternative future are being realized. The growing threat of cyber warfare, the demonstrated use of anti-satellite weaponry, and the continued expansion of terrorism on Earth, all mean that some of our implicit assumptions now need to be brought forth clearly – for the world to see and for potential adversaries to reckon with.

Space Policy needs to be more explicit. There are aspects of national space policy that have been implicit but not overtly stated. With the growing dependence of U.S. national security on activities in space, Policy needs to be updated to reflect the new realities of space operations. At a minimum, the following additions need to be considered for inclusion in the next public, formal statement of National Space Policy:

Space programs are a national treasure that must be nurtured and protected. It should be periodically and publicly reaffirmed that all elements of U.S. National Security Space are under the protection and care of the national security structure. Military, Intelligence Community (IC), civil and commercial components of NSS provide a special national strength that cannot be duplicated by any other means. This community continues to be perceived around the world as a source of national prestige, with obvious implications for global leadership.

Within these components, there are many overlapping communities, such as launch capability, launch facilities, satellite builders, payload houses, a diverse workforce, and a nation-wide educational community. Each community of the U.S. space program responds to different needs with essentially separate organizational structures. It is to our

Nation's benefit that we take proper care of this national treasure. We ignore that stricture at our peril.

Similarly, there should be an overt, stated, national policy intended to deter hostile acts against U.S. and friendly space systems. National policy must be clear that an attack on NSS forces, whether through kinetic, cyber or other non-kinetic means will be considered an attack on the U.S. homeland. International treaties support this proposition to a low level today. With the cyber threat growing daily, we need to emphasize cyber deterrence as part of the United States' determination to protect our national security space programs.

Domestic launch has national advantages. Twenty years ago, U.S. launch policy was based upon the assumption that it was possible to have a "free and fair market in which U.S. industry can compete." At that time the decision was made to continue the "use of U.S.-manufactured launch vehicles for launching U.S. Government satellites:

> "The long-term goal of the United States is a free and fair market in which U.S. industry can compete. To achieve this, a set of coordinated actions is needed for dealing with international competition in launch goods and services in a manner that is consistent with our nonproliferation and technology transfer objectives. These actions must address both the short-term (actions which will affect competitiveness over approximately the next ten years) and those which will have their principal effect in the longer term (i.e., after approximately the year 2000). In the near term, this includes trade agreements and enforcement of those agreements to limit unfair competition. It also includes the continued use of

> U.S.-manufactured launch vehicles for launching U.S. Government satellites."[8]
>
> The domestic launch industry took a hit with the 1994 National Space Transportation Policy: "With the end of the Cold War, it is important for the U.S. to be in a position to capitalize on foreign technologies - including Russian technologies - without, at the same time, becoming dependent on them. The policy allows the use of foreign components, technologies and (under certain conditions) foreign launch services, consistent with U.S. national security, foreign policy and commercial space guidelines in the policy."[9]

Today it has become painfully obvious that foreign subsidization of foreign launch vehicles has precluded any "free and fair market." Last year the GAO found that "The lack of a comprehensive U.S. national space launch strategy and unified oversight council is obstructing the domestic commercial launch industry's ability to grow and compete."[10]

It is past time to strengthen the National policy of preference for domestic launch. While there are significant obstacles to federal subsidies, incentives for and benefits of preference for domestic launch should be considered. A stronger domestic launch capability would drive innovation and inventions as it has done in the past, creating new companies and new investment in an American strength.

At the same time, in order to increase the range of options open in the event of an unexpected down-time for the Nation's fleet of launch vehicles, the U.S. should study the possibility of

[8] Commercial Space Launch Policy, Sep 1990
[9] National Space Transportation Policy, Aug 1994
[10] As reported by Jeffrey Hill, "GAO Report Highlights Issues with U.S. Commercial Launch Policy," Satellite TODAY, published by Access Intelligence, LLC, 4 Dec 2009

launching each of our space systems on foreign or commercial launchers. To the extent that cooperative programs can be pursued with the engagement of international launch agencies, the development of compatible launch, range and support capabilities would enhance the resiliency and durability of our launch capability. This study should be accomplished sooner rather than later, in order to understand and anticipate the vibration, thermal, acoustic shock and interface requirements of a complementary, supplementary, backup or contingency launch capability.

Junk in space is bad for everyone. Scientists and space engineers worldwide have long had concerns about the effects of orbital debris. NASA has sponsored debris studies for years, the Nation has a debris mitigation policy, and more and more companies are employing debris minimization techniques in their launch vehicles and satellites. Unfortunately, at the December 2009 DARPA / NASA-sponsored conference on Space Debris and Removal, the community recognized that the situation has passed a critical level with the Chinese ASAT event and the Iridium/Soyuz collision. The conclusions that were reached during the few days of briefings could be summarized as follows:

- Prior to the Iridium/Soyuz collision and the Chinese ASAT, the "big sky" theory dominated the space community. This theory supported the idea that space was so large, the odds were always in the favor of "no collision." Now, projections for the future have become scary and the big sky theory no longer describe the threats in orbit. We have moved into the arena of "Must Do Something."

- Policy must drive future mitigation actions to stop any deliberate creation of debris and enforce designing all rockets and spacecraft to create ZERO debris.
- Large rocket bodies and old spacecraft must be removed to ensure those large entities do not become future sources of additional debris through collisions with small stuff. If the objective of removing ten or more large bodies each year is implemented, the 2000+ major space debris pieces could be significantly reduced rapidly.
- Space Situational Awareness for the USG must be improved. This would include better sensors (resolution improvements by a factor of ten) and better computational processes to project positions and predict conjunctions.

> ..."every space-faring nation has a great incentive to avoid devastating wars in space that could greatly multiply the amount of debris in LEO."[11]

We Need to know what <u>else</u> is Up There. The Air Force has a mandate to track, categorize, and project conjunctions for all objects of value in Earth orbit. This policy of battlespace awareness and civil/commercial safety of flight leads to a demand that current capabilities cannot meet. Space Traffic Management (led by the U.S.) could be right around the corner if all the needs are understood. U.S. Strategic Command is well aware of this deficiency and has carried it on their Integrated Priority List as an unfunded requirement for several years. It

---

[11] "Space: The Highest Ground," Strategic Forecasting, Inc. (STRATFOR), October 19, 2009

is time to fund it and get on with it. National Policy should reinforce our determination to get a handle on this problem. Space Situational Awareness is one of the most critical improvements inside the NSS community that can be realized in the near term at a reasonable cost.

UNFINISHED DOCTRINE

Our Space Doctrine Needs Work. In contrast to Policy, Doctrine represents the sum total of all that we have learned and experienced about the "best" way to do things. According to Air University, "Military doctrine is what we believe about the best way to conduct military affairs."[12] From a doctrinal sense, space has often been called the "ultimate high ground." Today's military strategists are trying to accommodate the strength of the High Ground of Space, much as the great military minds of history, from Sun Tzu and Clausewitz to Napoleon, exploited the use of High Ground on Earth. At the same time, from a doctrinal sense, space power and airpower are frequently lumped in an arbitrary but undefined "aerospace power" that all too often only confuses both. In fact, the development of space doctrine today is about where Mitchell and Douhet were in developing airpower doctrine in the 1930s. For example, whereas airpower doctrine is frequently encapsulated in the comprehensive "Range, Speed, Maneuverability," space doctrine is almost diametric in its fundamental tenets.

---

[12]*Making Strategy: An Introduction to National Security Processes and Problems*, Chapter 11, Published 1988 by Air University Press. August 1988. pp. 163–174.

It is time to take a fresh look at Space Doctrine. In many cases, those who have worked on space programs over the years have been following space doctrine implicitly without consciously exploring its substance. It also occasionally happens that major space program decisions are made based upon budgetary or political factors that don't fully recognize their consequences in light of space doctrine. In addition, a space warfighter has different skill sets than a terrestrial warfighter, similar to the differences that separated the Army Air Corps and the Army in the 1940s. Accordingly, space doctrine needs to be written by people who understand space. The first step in understanding the nature of these differences is at the basic level of doctrine. Military strategists the world over have relied upon the Clausewitzian Principles of War for their doctrinal basis for operations on Earth [Objective, Offensive, Mass, Economy of Force, Maneuver, Unity of Command, Security, Surprise, Simplicity – see appendix B]. In the space domain, however, two other principles play significant roles.

Space Doctrine should include the Principle of Predictability. Space Operations are limited by the Principle of Predictability. Objects in space move in mathematically calculable orbits. Satellites are extremely predictable in their orbits. Any amateur astronomer (see www.heavens-above.com) knows exactly when a satellite will come into view, how long it will remain in view, and when it will go over the horizon and out of sight. From an offensive standpoint, the Principle of Predictability means that space forces are not very good for surprising an enemy. From a defensive standpoint, that predictability leaves satellites themselves inherently vulnerable to attack.

Space Doctrine should include the Principle of Instantaneity. Space Operations have the advantage of the Principle of Instantaneity. The sheer vastness of space, the inherent physical and electromagnetic nature of the vacuum of space, and the laws of physics that apply to objects in space mean that space is best suited for operations that can be done instantly – at the speed of light. This principle is not driven by a political decision, because the Principle of Instantaneity cannot be altered by the wave of a budget wand. It is not a tactical decision either, because the characteristics of space cannot be changed. It is not an economic decision that can be altered by throwing an unlimited amount of money at it. Movement of physical objects from one orbit in space to another takes longer, and is more complex, than most people realize. It uses energy that cannot be recovered or replaced, and laws of motion that do not fit within our Earthly frame of reference. Operations best suited for space – communications, navigation, weather monitoring, missile warning, ISR (Intelligence, Surveillance, Reconnaissance), directed energy, etc. – leverage the speed of light. "Movement" of zero-time-of-flight weapons between orbits is instantaneous, whereas physical movement between orbits is slow and expensive in terms of fuel consumed, opportunities lost, and energy spent.

Space is not a good medium for kinetic weapons. In space, bullets, missiles and bombs are too slow to get to a target, too predictable in their flight paths, too messy in terms of debris left behind, and too expansive from the standpoint of an Area of Responsibility (AOR). As we've seen recently, debris from kinetic events in space can remain in orbit for hundreds of years, posing a tangible threat to space operations for generations to come. In addition, the space "theater" of combat is staggeringly larger than a normal "AOR" in which air, land or sea forces operate. Kinetic weapons on Earth affect a small fraction of the ground below; kinetic weapons in space can scatter debris over thousands of miles in all directions, and stay in orbit, almost forever. Operational planners should apply space forces to tasks that require near-real-time solutions, and not to operations that require physical movement of objects, *e.g.*, "bombs from space."

Space professionals need to take a fresh look at what else is missing. In 1983, then-Lieutenant Colonel Roger DeKok and a small team of space experts developed the first Space Plan for the Air Force. That Plan defined the context in which we discuss space operations today, describing in detail the essence of four mission areas for space operations: Force Enhancement, Space Control, Force Application, and Space Support. Today, it is time to take an equally farsighted look at Space Doctrine, in preparation for sustaining the Nation's space forces for the next 20 years.

Chapter 3

SPACE STRATEGY – THE MISSION COMES FIRST

INTRODUCTION

The previous chapter shows how National Security Space Strategy must be consistent with National Space Policy and Operational Space Doctrine. Other aspects of Space Strategy, e.g., concept of operations, organizational considerations, training requirements, etc., flow from the Strategy. However, once Policy and Doctrine are understood, the next step in developing Strategy is knowing what the Mission is.

MISSION

All National Security Operations Start with a Mission. All else follows. A clear understanding of the missions supported by space forces is absolutely essential when such operations affect the lives of our soldiers and the well-being of our Nation.

Space forces support virtually every USG Mission. At the highest level of national strategy, space forces have a similar geographic responsibility as forces on land, at sea and in the air:

- Help our National leadership understand the global and tactical situation.
- Help provide the enemy's intentions to National leadership.
- Help provide global situational awareness in space, in the air, on land and at sea.
- Support the conduct of U.S. and allied military operations anywhere in the world.
- Support global U.S. civil and commercial ventures.

Within the space domain, the overarching mission of space forces is to bring the full force of space power to bear. Depending upon the circumstances involved, this top-level mission could include several responsibilities.

- Ensure U.S. access to space.
- Protect U.S. and allies' satellite operations, including commercial and civil satellites.
- When directed by National leadership, deny an enemy the benefits of space.

The Mission, however, is directly related to the Threat.

THREAT

Space is a full spectrum threat environment - and not just from the bad guys. Similar to submarines when submerged, satellites are always at risk, even from the "natural" environment. The difference is that for objects in space, there is no safe "home port," no "lee of the island" where satellites can go. Threats to satellites and their supporting ground structure range from the embarrassingly simple (a car ramming a microwave tower or a backhoe cutting an underground fiber optic cable) to the profoundly complex (a direct-ascent or co-orbital anti-satellite weapon). Satellites are at the mercy of a wide variety of natural hazards and hostile threats such as kinetic (collision) as well as non-kinetic (electronic jamming or cyber manipulation) warfare. Electronically, at a very generic level, satellites can be disrupted by solar flares, jammed by radio transmitters, blinded by lasers, crippled by cyber attacks, or neutralized by other "non-kinetic" means. Physically, satellites can be damaged or destroyed by nature (meteorites or space debris) or kinetic attacks. There are probably other threats as well, but for obvious reasons, satellite manufacturers are reluctant to discuss the vulnerabilities of their systems. In any event, building comprehensive, detailed catalogues of potential threats is a manpower-intensive process.

> ..."attack from a direct-ascent satellite-killer is not the only danger out there. Threats range from proliferating space junk and orbital vehicle collisions to ground-based jamming, which now can be used to neutralize or disrupt spacecraft in medium Earth orbits."[13]

---

[13] Rebecca Grant, "Insecurity in Space," *Air Force Magazine*, October, 2009

Launch itself is a very hostile regime. A satellite undergoes extreme shaking, rattling and rolling as it is launched. Acoustic, vibration, and gravitational forces are all severe effects that a satellite must be designed to withstand.

> "The only natural predator of a spacecraft is the rocket."[14]

Space is deadly. Once in orbit, satellites have to operate in a very hostile environment – vacuum, extreme cold, radiation, weightlessness, etc. Designing and building satellites with the durability to operate in such an environment is very expensive and absolutely necessary because there is no easy way to repair satellites in higher orbits. Without a "10,000 to 23,000-mile long screwdriver," satellites are burdened with an "out-of-reach" characteristic that adds exorbitantly to the total cost of each system.

Man-made Threats abound. For the purpose of developing National Security Space Strategy, man-made threats can be categorized by the level of technological sophistication required to be able to carry them out.

In-your-face attacks are hard to do and hard to prevent. While the ability to rendezvous in space is the ultimate step for achieving proficiency in human spaceflight, it is also a skill that can lead to the co-orbiting, or collision, with another satellite. Designing, building and deploying a rendezvous capability is an expensive proposition requiring very sophisticated technology; only a few nations in the world can actually afford the high entry fee to be a player at this

---

[14] Attributed to Col.(Ret.) Jim Mannen

level. This is a skill that only peer adversaries have and practice, most recently demonstrated by the Chinese ASAT. Defending against it would be very expensive, which leads to a tradeoff in deciding whether the likelihood of such an attack is worth the high cost of building an adequate defense against it.

Satellites can be attacked by forces on Earth. Simply tracking and interfering with a satellite above the Earth requires a more modest entry fee and can be conducted by a wider range of opponents. This capability requires access to tracking radars and/or optical systems that can provide significant counter-satellite capabilities. This threat includes (1) hiding assets when U.S. satellites are scheduled to pass overhead, (2) radiation or laser interference with mission capability, or (3) radiation or laser destruction of mission payload or spacecraft.

Ground facilities on Earth can be attacked. At the low end of the spectrum, and therefore much more likely, are "conventional" military or terrorist attacks on ground facilities supporting satellite operations. This capability exists even in the hands of Somali pirates. In an open society this threat is very real and probably the most likely to be employed by the widest range of actors. Weapons could include hand grenades over the fence and into a communications antenna, cutting the power source to a ground facility, brute force jamming of satellite command and control, or even cyber attacks against the network supporting satellite operations.

Our defense against hostile threats is limited. Historically, the USG approach to defending space operations has been limited to peripheral factors.

We invoke National Sovereignty. In the spirit of deterring attacks on U.S. and friendly space forces, the U.S. has often stated that U.S. satellites are national assets and any approach or interference will be considered "saber rattling" and will be taken seriously.

We try to Identify and Describe ("Characterize") the Threat. The USG uses a variety of sensors and intelligence collection processes for developing "Space Situational Awareness" and determining the intent of potentially hostile systems. For some DoD programs, satellites carry sensors that can characterize attacks against the satellite itself.

We design for survivability in the Natural Environment. Satellites are designed to operate in, and survive the rigors of, a hostile - but benign - environment in space. For the most critical satellites, additional protective measures are included in their design. Milstar and Advanced EHF, for example, carry additional protection because they are designed to provide strategic communications across the spectrum of warfare. For less critical satellites, specific protective measures to counter hostile attacks are generally too expensive and are not included.

We have demonstrated a deterrence capability if needed. On at least two occasions, the U.S. Government has shown that it has the ability to respond to a provocative act in space. Several years ago, the Air Force developed and launched an F-15-carried ASAT to counter perceived threats from the Soviet Union. More recently, the Navy shot down a disabled (but friendly) satellite in low earth orbit, to reduce the possibility of causing a hazardous debris problem near population on Earth when the satellite de-orbited on its own.

Both were demonstrations of the technology needed to respond if deterrence failed.

The Threat is getting more real every day. In terms of an improved National Security Space Strategy, the logical question to ask is, given the threat of a shooting war in space sometime in the future, is now the time to move defensive (protective) or offensive (strike) weapons into space? The answer to this question, of course, must recognize that other nations would be expected to respond in kind, leading to an escalation of the likelihood of armed conflict in space.

At the same time, there is a large contingent of policy and doctrinal leadership in the United States who argue that, as a 21st Century environment, space could leverage the treaty route used in the international arena. The 1967 Space Treaty established "norms" that are followed today such as no weapons of mass destruction and nations are responsible for damage and registry. The concept of "no weapons in space" is not unrealistic if one believes that humanity has learned something from the 20th Century, with its modest steps toward reducing the threat of weapons of mass destruction, and its futuristic hopes of moving humanity to other worlds. If you combine the best of global space activities, space exemplifies the 21st Century paradigm of global communities working together (e.g., global warming, disaster relief, NATO expansion with Warsaw Pact countries, etc.), and the recognition that space has unique strengths and characteristics that can benefit all mankind.

In either case, the over-arching National Security Space Mission remains the same: **Protect and Serve.** Encourage expansion in space by all nations while enabling the U.S. to operate freely.

CONCEPT OF OPERATIONS (CONOPS)

The Concept of Operations (CONOPS) is the Backbone of National Security Strategy. CONOPS describes the "how" in "how to accomplish the Mission." CONOPS addresses forces, troops, materiel and employment plans – in short, every aspect of what will be necessary to ensure success. As important as a CONOPS is, it turns out that there is no commonly agreed-upon definition of what specifically must be included in order to have a "complete" CONOPS. In fact, no two military Services – or even military organizations within a single Service – agree on the definition of a "complete" CONOPS. There is even less agreement between the Intelligence Community and the Defense Department. Indeed, there is no explicit, comprehensive, over-arching, cross-Agency CONOPS for space programs supporting missions of both the Intelligence Community and the DoD. There are even fewer ties with civil and commercial space enterprises. Therefore, we are using the CONOPS construct developed by Air Force Space Command's Independent Strategic Assessment Group (ISAG). The ISAG defines a CONOPS by answering nine key questions:

- What are the **missions** to be accomplished? There are many missions, some with common characteristics. The Joint Space Operations Center (JSpOC) calls out Mission "threads," that include things like launch, space surveillance, communications, etc. The Intelligence Community (IC) includes strategic indications and warning. Civil communications, weather (military and civil), and NASA missions fill out the portfolio.
- What is the **Combatant Commander's (or Intelligence Community leader's)** intent? For example, provide uninterrupted communications to military forces

deployed in a particular area. Gather information that would indicate the intent of a potential adversary's moving a large number of forces closer to an area where there are known opportunities for mischief.

- Who are the **key players**? Players are involved at all levels of the Executive Branch, and its many Departments and Agencies. The "key" players are usually limited to those in the direct chain of command, as well as specific users and forces supported. For the commercial world, corporation leadership fills this role.
- What are the **command relationships** and organizational interfaces between the key players? This question goes to the heart of, "Who is in charge?" It is particularly relevant when a satellite is tasked to do more than it is capable of doing, and priorities need to be set regarding which requirements are satisfied first, and in what order. The "forces supporting" must be addressed, as well as the "forces supported."
- What **resources** (materiel capabilities) of the system-of-systems are used to execute missions? Communications, C2, power, logistics, supply, transportation – whatever is needed to accomplish a particular mission.
- What high-level **operational tasks** are required? These are the top-level tasks that are identified by looking at the Mission and seeing what principal activities are required in order to accomplish it.
- What **connectivity** is required between the participating system-of-systems elements? Connectivity, communications – the links from requirements-tasking-collection-processing-exploitation-feedback cycle, looping back to requirements and starting the cycle again.

- What are the **training** tasks?
- How is **mission readiness** determined and reported?

Once a CONOPS is established, the supporting units must Organize and Train for the Mission. Organizing and training both directly affect whether a National Security Space Strategy is realistic or not. Let's discuss Organization next.

ORGANIZING FOR SPACE

Is Space Organized Correctly? The Defense Department generally treats Organization as one of the seven basic elements of the DOTMLPF[15] paradigm. Because organizational changes are relatively easy to make, they are an attractive option for giving the appearance of "progress" or "improvement."

> "We trained very hard...but it seemed that every time we were beginning to form into a team we would be reorganized. I was to learn later in life that we tend to meet any new situation by reorganizing; and a wonderful method it can be for creating the illusion of progress while producing confusion, inefficiency and demoralization." ...Allegedly written in 210 BC by Petronius the Arbiter (there is some doubt as to its authenticity, but little doubt as to its truth)

[15] Defined in the Joint Capabilities Integration Development System (JCIDS) process as Doctrine, Organization, Training, Materiel, Leadership and Education, Personnel, and Facilities

Many have suggested that space operations would be better served if space forces were a separate military Service. It is by no means clear that the overhead burden of a separate Service would offset the advantages of a space-focused career path. But now is certainly the time to think about whether the U.S. is organized properly for a real Space War. At the "Space Commanders' Forum" session during the 2009 Space Symposium, speakers talked about a desire not to repeat our lesson from 9/11, where it took an attack to get us to organize efficiently for homeland defense. It would be far better to organize for space warfare before it happens, and sooner rather than later.

In any event, no organization is so imperfect that good people can't make it work. What is important is providing top-down support, good leadership and "just enough" funding, all the while allowing good people to do their jobs.

Some Organizational principles have withstood the test of time. Any organization that is responsible for life-or-death matters should be built upon the following organizational principles:

- **National Security organizations must focus on the Mission.** If an organization doesn't start with the Mission, you're asking for trouble. You must also keep the Mission simple and easy to understand. An example: Troops deployed to Bosnia had their Mission and Rules of Engagement printed on a 5" x 9" placard that they carried with them
- **The Organization must fit the CONOPS.** Organizations should ideally be structured by the CONOPS. Insist on a clear - and agreed-upon - CONOPS before trying to organize.

- **The Organization must have a clear Chain of Command.** Once a CONOPS is developed and agreed upon, a Chain of Command should be established that provides two essential imperatives:
    - **The Chain of Command must provide clear lines of responsibility.** Historically, this has been a bit of a problem for space forces, particularly on the acquisition side. For many years the Air Force, at least tacitly, assumed the role of "Executive Agent" for DoD space operations. That role is gradually changing: OSD recently took management responsibility away from the Air Force for a few space programs, and moved ACAT (Acquisition Category) responsibilities for selected high-value space programs to OSD. Neither of these actions is very helpful in defining clear lines of responsibility for space operations.[16]

      The latest review of organizational constructs comes from the Obering Panel, preliminary results of which were published last September. While the details of the Panel's findings remain to be published, it is not intuitively obvious that changing organizational structure at the highest levels improves either clear responsibilities, or clear accountability as discussed below. For example, recent recommendations that DoD re-combine the Air

[16] The Defense Acquisition University (DAU) describes the ACAT ranking process as one in which size, complexity, and risk generally determine the category of an acquisition program. Programs are designated as either an Acquisition Category (ACAT) or, for smaller programs, an Abbreviated Acquisition Program (AAP). The Milestone Decision Authority (MDA) designates Major Defense Acquisition Programs as ACAT I if they are high-value programs whose funding exceeds either RDT&E or Procurement funding thresholds. Programs designated ACAT II are still major systems but they do not meet the criteria for ACAT I. A weapon not designated ACAT I or II but that involves combat capability will normally be designated ACAT III. ACAT IV is assigned to smaller ACAT programs.

Force Under Secretary and NRO Director positions would restore at least the effect of a cleaner chain of command. Incorporation of the Panel's findings, then, must be contingent upon the degree to which these Measures of Effectiveness (MOEs) are satisfied.

- **The Chain of Command must provide clear accountability.** Successes – good performance – should be rewarded; failures – poor performance – should be recognized and steps taken for improvement. Personal, corporate, and government accountability is important. At the personal level, assignment choices and career development plans should be consistent with accountability metrics. At the corporate level, Award Fee contracts have historically provided a level of accountability. At the government level, accountability is essential for maintaining a high quality of performance. Recently the DoD has taken prominent steps to restore a sense of accountability in acquisition programs. Long overdue.

Measures Of Effectiveness (Moe)

Measures of Effectiveness are intended to provide constructive, definitive indicators of performance. In this section on Organization, and in following sections on Training, Acquisition, Operations, and Sustainment, the MOEs listed constitute a checklist of metrics for grading performance. These MOEs can, in effect, be a scorecard that can be administered from within, or by an outside agent. In either case, MOEs are only as useful as leadership allows them to be.

Measures Of Effectiveness (Moe) – Organization

Is the **Acquisition** organization structured to…

- …reward success and learn from failure?
- …encourage a healthy Industrial Base of prime contractors, subcontractors, and third-party vendors?
- …foster a robust Technology Base?
- …respond quickly to customer demands ("Speed to Need")?
- …support, develop and enforce clear metrics for measuring performance and achievement?
- …enable aggressive Risk Reduction through healthy Research and Development (R&D)?

Is the **Operations** organization structured to…

- …respond quickly to changing operational requirements?
- …adapt quickly to technology breakthroughs, e.g., stealth UAV?
- …encourage a "Multiple Security Level" (as opposed to a Multi-Level Security) environment for ease of use with Allies and Coalition Forces?

Is the **Sustainment** organization structured to…

- …provide continuity in staffing?
- …document, protect and disseminate Lessons Learned in order to avoid "reinventing the wheel"?
- …implement Learning Curve efficiencies and cost reductions?
- …transition the "Marching Army" to other requirements efficiently?

SPACE TRAINING

In acquisition, operations and sustainment, Training makes the difference between success and failure. Training is so important that it is another of the seven basic elements of the DOTMLPF paradigm. Training for space forces is a relatively structured process that has evolved as USG space forces have grown. As with other national security disciplines, training for an acquisition career is not the same as training for an operations career, and both are separate and distinct from training for sustainment.

Today, space training includes everything from the art of warfare and the physics of space, to an in-depth understanding of operational space doctrine, as well as high-technology satellite production management efficiencies. While there will always be a shortfall of funding and resources in any training activity, the space profession has three specific areas where Training needs more attention:

- Providing in-depth training for the space acquisition career field that can only come with long-term exposure to the acquisition profession, as well as to the nuances of buying high-value, low-quantity, state-of-the-art space systems.
- Keeping Training current in an environment where technology is leading force structure.
- Avoiding "look good in the shower" Training programs that don't live up to expectations.

Measures of Effectiveness (MOE) – Training

Does the **Training**...

- ...ensure efficient operations despite the frequent turnover associated with most military positions?
- ...attract and build a quality force and promote recruitment and retention?
- ...facilitate career progression?

Chapter 4

NSS ACQUISITION STRATEGY – LEAD BETTER, FOLLOW WELL, BUY SMART

INTRODUCTION

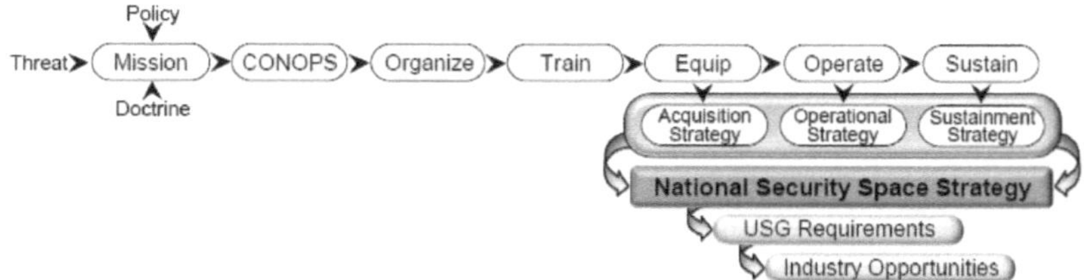

Space Acquisition is a lot harder than it looks. Acquisition of space systems is an extraordinarily difficult task. Libraries have been filled to overflowing with articles describing how the acquisition of space systems is irreversibly broken, and that government and industry leadership has sorely mismanaged the task of building space programs that serve and protect our Nation, our Allies, and our friends. Oddly enough, you won't find nearly the volume of articles describing how well these same space systems have accomplished incredible achievements. The truth, no doubt, lies somewhere in between. It is safe to say that extraordinarily dedicated people have led industry and government teams in accomplishing feats that fit somewhere in the "Wonders of the World" top ten. It is equally safe to say that the space acquisition process is no more broken than other acquisition processes that have tried to save money by trimming a force structure that took decades to build. The difference with space programs is that the high unit costs draw more attention. Space systems are expensive because every aspect of their operation is a challenge unto itself.

The "Laws of Physics" make getting to orbit a wonder in itself. Designing a satellite – and a launch vehicle that will put it in orbit – is hard from the very start. It takes a lot of energy to get a satellite high enough into space, and moving fast enough, that it won't fall back to Earth. At the lowest efficient orbit that can be sustained without using additional power (about 300 miles above the Earth), a satellite has to be traveling at 17,600 mph. It doesn't matter how big the satellite is, what it does, or how much it weighs, it is very difficult just designing and building a launch system that will make anything go that fast. It doesn't help that most of what is sitting on the ground just before launch never makes it to orbit. In fact, only 5% of the mass on the launch pad ends up as an operational satellite in Low Earth Orbit (LEO). The fact that we can beat gravity almost routinely today is a lot more incredible than most people realize.

Extraordinarily complex systems require exquisite Systems Engineering. Building satellites is a complex engineering problem. They are very hard to build. In many respects, satellite builders have more demanding performance criteria than those for aircraft, ships, submarines or ground vehicles, especially in the areas of mean time between failure (MTBF), mean time to repair (MTTR), lifetime, mean mission duration (MMD), self-diagnosis-and-repair, and environmental threats. Because they can't be brought back to Earth for routine or periodic maintenance (a "3,000 mile oil change," for example), they have to be designed to be maintenance-free for a very long time. Much longer than we design for most Earth systems. Satellite development costs significantly change the paradigm for program development. In addition, the operators of these complex systems require technical savvy throughout the operation's chain leading to extensive training and valued employees you cannot afford to rotate too often. All of these technical complexity factors mandate a comprehensive (and expensive) Systems Engineering capability.

Development costs can't be shared across a long production line. One reason the per-unit cost of a satellite is much higher than the per-unit cost of other major weapons systems on Earth is because there are so few of any one kind of satellite in space. GPS has fewer than 30 operational satellites on orbit; Iridium launched only 92 commercial satellites. Most other constellations are under half a dozen. This means that the non-recurring (NRE) development cost for a new program can't be amortized over several thousand production items, which would otherwise bring the per-unit cost down.

Historically, recommendations to improve acquisition have been consistent. Early in the development of national security space programs, Colonel Lee Battle laid down his laws for space acquisition. His guidelines focused on getting the right people, working closely with industry, and taking personal responsibility for results.

Battle's Laws[17]

1. Keep the program office small and quick reacting at all costs
2. Exercise extreme care in selecting people, then rely heavily on their personal abilities
3. Make the greatest possible use of SSD [SMC in those days] supporting organizations. You have to make unreasonable demands to make sure of this support.
4. Cut out all unnecessary paperwork
5. Control the Contractor by personal contact. Each man in the program office has a particular set of Contractor contacts
6. Hit all flight and checkout failures hard. A fault uncorrected now will come back to haunt you.
7. Rely strongly on Contractor technical recommendations, once the program office has performed its function of making sure the Contractor has given the problem sufficient effort
8. Don't over communicate with higher headquarters
9. Don't make a Federal case out of it if your fiscal budget seems too low. These matters usually take care of themselves
10. Don't look back, History never repeats itself.

[17] McDonald, Robert A., ed., Corona Between the Sun and the Earth, The First NRO Reconnaissance Eye in Space (Bethesda, MD: American Society for Photogrammetry and Remote Sensing, 1997).

Former Secretary of Defense David Packard's guidelines were even shorter: Hire the best people for the job, give them the authority and resources they need to do it, and then get out of their way. Dr. Robert Naka, with many years in the "black" (classified) and "white" (unclassified) space world at the highest levels, recognized that the classified approach to systems acquisition had some major strengths. His main thrusts were: (1) significantly less oversight leading to fewer briefings, and (2) technical team stayed with the projects for many years providing continuity.

During the 2009 Air Force Association "Air & Space Conference," Dr. Paul Kaminski (former Under Secretary of Defense (Acquisition, Technology and Logistics)) commented that, "What's needed is a focus on skilled people who are given the authority and money to run programs as they see fit."[18] He suggested five steps to fix today's acquisition melt-downs:

> **First,** system engineering and program development planning require people with "domain experience," practical knowledge of the area in which their product will function, to better understand how to organize and schedule new projects, and more money should be put in up-front to map out the course of programs.
>
> **Second**, the Pentagon needs to "align program responsibility, authority, and accountability," so that managers aren't serving conflicting masters and requirements. Managers should also be given greater authority to manage their own programs, be allowed to adjust schedules, and be given a 10- to 20-percent management reserve to cope with the inevitable problem of developing new technology.

---

[18] Tirpak, John A., "The Doctor's Prescription," Air Force Magazine.com, September 14, 2009

Third, programs need funding stability, and Kaminski said both Congress and the Pentagon need to give up some of their "flexibility" to tinker with funding so that cash flow and program burps can be handled.

Fourth, there must be "early and serious attention" to test and evaluation. How programs will be judged and proven needs to be part of the contract and needs to be adequately funded.

Fifth, Kaminski said the development time of programs must be shortened from 15-20 years to five or less. Lengthy acquisitions have created a generation of program managers who have worked on only one program – or less – in their careers. Success in future defense technology hinges on growing managers who have experience in several programs and from the component level up to a full system, Kaminski said.[19]

Ten Rules for Common Sense Space Acquisition.[20] Major General Tom Taverney (USAF, Ret.) and Colonel Jim Rendleman (USAF, Ret.) applied their many years of experience in acquisition and developed their own basic rules. It should come as no surprise that they parallel Battle's Laws and implement Dr. Kaminski's approach.

1. Put together the right team – one that is small, agile, intellectually honest, and quick to respond; this team is the foundation for success.

---

[19] *"Washington Watch," Air Force Magazine, November, 2009. Pg 10.*

2. Execute, or suffer execution.
3. Establish a solid baseline.
4. Control the baseline; it is your lifeblood.
5. Manage risk; it never goes away on its own.
6. Make the program schedule a leading metric.
7. Nip problems in the bud; they usually do not get better with time.
8. Test and verify; one test is worth a thousand opinions.
9. Communicate; it is more important than organizing.
10. Deliver; it is all about delivering the needed capability to the user.

Space Acquisition is a very unyielding process. In spite of consistent recommendations by these and other space notables, we seem to be making slow, if any, progress even though improvements could produce significant benefits. There is little doubt that it will take a full-court press across all aspects of Acquisition to make headway. The USG must buy and build smarter by re-balancing risk and rewards in order to incentivize the performance that it wants from government buyers as well as industry suppliers. In order to achieve these objectives, this monographs recommends a three-prong strategy for improving space acquisition: **Lead Better, Follow Well, Buy Smart.**

[20] Ten Rules for Common Sense Space Acquisition. T. D. Taverney and J. D. Rendleman. High Frontier, USAF Space Command, Vol 6. # 1. Pg 53 – 65.

LEAD BETTER

Good Leadership is central for the success of any strategy. "Leadership and Education" form one of the seven basic elements of the DOTMLPF paradigm. Characteristics of good leadership are probably universal, and good leaders have filled a range of roles, from commander to manager, advocate, supporter, mentor, mover/shaker, cheerleader, accountant, caretaker, janitor, and more. Every person brings different qualifications and experience to a leadership role. In the space profession, desirable attributes vary widely because of the diversity of roles. At a minimum, operational experience, acquisition experience, and people skills are all desirable. The key is relevant experience. The more inexperienced a water-walker is (such as a new test director, project officer, program manager, squadron, group or wing commander), the less qualified that person is in the domain. In addition, they have a very steep learning curve and a greater likelihood of a failure somewhere along the way. Finding the right leader for each unique situation is a constant challenge that needs continued attention. The leader of a space launch unit should be knowledgeable in the technology and its associated risk levels. An example can show the differences in perspective required:

> **Launch Window (minutes) for B-52 Wing**: Operational commanders of a B-52 wing will move up the ranks if they can ensure launch windows of 2 to 3 minutes for separate takeoffs of individual aircraft for a major Operational Readiness Inspection.
>
> Or
>
> **Launch Windows (weeks to months) for space launch**: Operational commanders of the launch

wings will be successful if satellites get to the appropriate locations in orbit. Launch windows vary depending on orbital characteristics. Many have weeks or months-long launch windows, or in the case of a MILSTAR launch, 18 months spent on the pad.

In the space arena, where most systems count from one to five satellites (of course GPS is an exception), the care and feeding of satellites consumes an inordinate amount of time and energy. Commanders must understand the differences of space, as they do not have depot level maintenance and spare jets parked on the flight line.

FOLLOW WELL

The larger any task is, the more important the "followership" role becomes. Buying new space programs is every bit as demanding as outfitting a new aircraft carrier or deploying a new aircraft. All involve an incredible degree of interaction, coordination and followership. The key is recognizing that making our National Security stronger is a noble effort. We are all in this together, and teamwork is essential if we are to work toward that common goal.

BUY SMART

In order to strengthen National Security Space acquisition, this monograph recommends the following five strategies.

1. The USG should prioritize program decisions based upon criticality. First provide the satellites that support the Nation's highest priorities. Price and schedule are critical, but protection of the Nation is paramount. This **is** rocket science – higher risk and higher cost are unavoidable, should be anticipated, and must be managed rigorously. Programs that don't make the cut should be terminated.

- **Protect the Crown Jewels First.** Follow the principle of proportional defense. Critical systems must get more attention than non-critical systems, in order to ensure their ability to accomplish the mission. Backup ground stations can reduce ground station vulnerability. Alternate communication paths and backup cross-links can improve communications reliability. Increased redundancy in design can minimize the effects of single-event upsets in satellite electronics. Critical systems must have sufficient protection for as long as it takes to replace them. Critical systems must be able to degrade gracefully in order to give decision-makers time to provide replacements.

  The abundance of potential targets in space mirrors the complexity of the problem facing the electronics world of the Air Force. Electronic Systems Center Commander LtGen Ted Bowlds admitted that protecting the entire Air Force network is practically impossible. "Maybe we don't have to protect everything," he suggested. "Maybe we ought to decide what the crown jewels are and focus on protecting them."[21]

---

[21] Paone, Chuck. "Leaders call for balance between network use and protection," 66th Air Base Wing Public Affairs, Air Force Materiel Command, reporting on comments by Air

- **Build "Assured Launch" at a bargain.** One of the most pervasive - and misleading - rallying cries in the National Security Space business today is the call for "Assured Launch." This is the asserted requirement that a high-priority user must be able to count on having a rocket available to launch their payload when they need it. What is conveniently overlooked is that the rocket is rarely the cause of delaying the availability of a new satellite. In fact, if you look at all the satellites being built today, and all those projected for deployment in the next ten years, there are plenty of rocket designs available to launch them when needed. A more complete description of this shell game is provided in the RAND Evolved Expendable Launch Vehicle (EELV) report. The bottom line is that there are enough small and medium rockets available today to accommodate launch requirements for smaller satellites. For larger satellites, EELV has proven to be adequate for identified requirements for the next several years. One recommendation that would make the biggest improvement in availability calculations would be to "build one EELV ahead" in the production line. Having a spare EELV on hand would virtually guarantee that no satellite would ever be in the untenable position of "waiting for a ride."[22]
- **Robust with Hosted Payloads.** The concept behind "hosting" payloads is to put a mission payload on a satellite that is not part of that mission family. This is a

---

Force Electronic Systems Center Commander Lt. Gen. Ted Bowlds at the 7th annual Net-centric Operations Conference, 24 Sep 2009 in New Castle NH.

[22] "National Security Space Launch Report," The Congressionally Mandated National Security Space Launch Requirements Panel, RAND National Defense Research Institute, 2006.

good way to take advantage of excess capacity on some satellites. For example, the addition of a GPS-like navigation package to a communications satellite in geosynchronous orbit has improved GPS navigation accuracy by adding a non-traditional orbit to position calculation.

Hosted Payloads can also improve constellation survivability because they would require an adversary to attack more satellites in order to defeat a particular mission. Beyond that, for some applications, hosted payloads can be a quicker or less expensive way to get to orbit without having to build an entire satellite to test an experimental prototype. Designing host satellites with the ability to accept ride-alongs late in production makes this feature much more attractive and allows hosts to offset some costs. Of course, there are always potential downsides of working across stovepipes. In addition to the usual "not invented here" obstructionists, using hosted payloads requires coordination of Command and Control (C2) and data handling systems across constellations. Another operational consideration is resolving the rules if either the host or the secondary payload were to have an operational problem that requires additional fuel or power to fix, which could come at the expense of other on-board systems. Most satellite builders are loathe to let anyone else ride-share because adjudicating operational issues tends to be "just too hard." That is why it is very important for hosted payloads to coordinate a detailed "prenuptial agreement" with their host well before launch.

2. **USG acquisition centers should go "Back to Basics" – and stay there.** In keeping with SMC's rallying cry, the USG should develop an unmatched expertise at managing what really matters:

- **Cost, schedule, performance, and risk remain the cornerstones of acquisition.**
- **The USG should compete more.** Developers and acquirers of space systems must not forget the basics of procurement
    - Contract for as much competition in <u>every</u> aspect of the space force structure as possible, as early as possible, and for as long as possible.
    - Remember that incumbent expertise must be a significant factor in this competitive environment.
    - Document performance in award fees, CPARs and other metrics.
- **Beat Murphy's Law and think <u>smaller</u> together.** One way to make the staggering complexity of today's systems more manageable is to break them into smaller pieces that can be evaluated and procured independently. This will simplify the management steps and still produce a capable, complex system when putting the parts together. The software world has adopted this approach over the past several years, with considerable success. Industry and Government should sub-divide complex systems into more manageable pieces.
- **Make WIFMs count and buy risk down <u>cooperatively</u>.** Technology can lead to tremendous leaps in mission capability. However, the risk buy-

downs must be accomplished early and across mission domains. Both industry and USG must recognize these opportunities and work aggressively to improve the Technology Readiness Level (TRL) of promising technologies. Most organizations and individuals make choices based upon WIFMs: What's in It For Me? Industry and Government must develop risk reduction incentives that appeal to WIFMs and create win-win situations.

3. The USG should work toward a <u>balance</u> in acquisition. The operative word here is "balance." Historically, acquisition re-directions have been from one extreme to another. In an attempt to fix one mistake, we all too frequently create another. And of course, for each of these "balance" equations, the key is having the experience to be able to determine where the fulcrum is.

- **Balance the Mission Assurance Chain.** Identify the end-to-end system-of-systems architecture that leads to Mission Assurance, and anticipate the "weak links in the chain." Where alternate systems provide at least a modicum of Mission Assurance <u>do not spend</u> a lot of money on increasing a satellite's ability to survive a hostile attack. In a crisis, civil weather satellites can cover for DMSP. Commercial comsats (that are usually the prime anyway) can back up military comsats. GPS satellites provide resiliency by their sheer numbers in orbit. Mission Assurance should take into account the contributions of <u>all</u> systems to the mission, not just the contribution of specific military satellite constellations.
- **Balance "Technology Push" and "Requirements Pull."** It is not one or the other. Success depends upon

having both. An effective strategy requires that the process recognizes two strengths: (1) the "futures" experience of a professional technologist with space credentials who can comprehend what might be needed in 40 years (15 years to develop after the program has been approved, plus 25 years of the system's lifetime), and (2) the "current ops" experience of professional Intelligence Community or DoD operators who understand the near-term needs (three months to three years). Each of these strengths must be blended to ensure the proper perspective when approving a new program.

"If we had waited for Requirements Pull, we wouldn't have the systems that we have now."[23]
Mr. Jimmie D. Hill, former Director, NRO.

- **Balance "Incremental Development" and "Block Changes."** It is not one or the other. The choice will always depend on where the program is, what the latest requirements are, what technologies can be applied, and who is in charge.

4. The USG should design for the long haul. Given the extended lifetime of most satellites, the smart move is to look ahead and pave the way for a stronger Sustainment capability.

"The cheapest satellite is the one that is already up there."[24] *Pete Aldridge.*

---

[23] Attributed to Mr. Jimmie D. Hill, former Director, National Reconnaissance Office
[24] Attributed to Hon Pete Aldridge, former Under Secretary of Defense (AT&L)

- **Break Stovepipes.** The biggest improvement that can be made in operational responsiveness today involves how data from space is handled once on Earth. Breaking stovepipes will go a lot further toward saving soldiers' lives than any other single investment in NSS forces. We should step up to this travesty before it is too late. Mandate cross-stovepipe fusion for all NSS missions, including Joint and Allied interoperability. Improve the way we use the satellites we already have. There is a significantly greater return on investment (ROI) here than from buying new programs – not a popular consideration in aerospace industry circles, but the truth. Incentivize cooperative tasking, stovepipe crossover, and data fusion. We will never make progress in data fusion and data interoperability until we step up to paying for it – even if that means at the expense of the sources. Moving dollars is probably the only way to put the focus of attention back on getting the biggest bang for the buck in terms of making the biggest near-term difference in support to the warfighter. Operators are going to have to make this case, because without operational dissatisfaction with the *status quo*, nothing is going to happen to upset the Acquisition apple cart.

- **Design to Operate Transparently.** Space forces must be fully compatible with their complementary systems on Earth. The user should not have to care whether a particular service is being provided from space or from an equivalent system on Earth. Transparent operations are usually facilitated when common standards and interfaces are used. Unfortunately, commonality is difficult to implement when proprietary designs control the baseline.

- **Reward Net-Centricity.** In the old days, the government typically wrote contracts that rewarded companies for meeting or exceeding the performance specifications. That worked until cost and schedule fell by the wayside in a determination to squeeze out the last ounce of performance. So contracts added cost and schedule as metrics for success. Again, that worked for a while. Then programs were designed to meet the three goals but at a higher risk. So today we have the "cost, schedule, performance, risk" paradigm, and now that isn't getting us to where we need to go. Companies are determined to keep their designs proprietary and their engineers challenged, which means that few systems are truly meeting USG requirements because they won't work with other systems on the battlefield. Next step? Add "net-centricity" as a performance measure of merit, and put money behind it. It's one way to force the fusion the operational user needs.
- **Build Extended-Lifetime Reliability in Up Front.** Given the inherent difficulty of getting satellites into space in the first place, it is usually more cost-effective to launch a well-built, reliable satellite that will last a long time, rather than to try to launch several cheaper satellites that don't have the same expected lifetime or mission reach. In other words, the launch step in the "Probability of Success" (Ps) chain is a significant part of the equation – similar to the cost of a Navy Carrier Battle Group used to bring aircraft to the fight.

  "More than a quarter-century after a hard-luck launch aboard the shuttle Challenger, the pioneer of NASA's constellation of tracking and communications satellites is being retired from service." A failure on the TDRS 1 satellite "led NASA officials to finally retire

TDRS 1 after operating almost four times longer than designed. Officials can still command the satellite, but it is now useless to customers." Described as the "queen of our fleet," by Roger Flaherty, Space Network project manager at the Goddard Space Flight Center, the spacecraft will be moved out of its geosynchronous orbit for final decommissioning. "Two new TDRSS craft are being developed for launch in 2012 and 2013."[25] ...

5. The USG should intensify ORS. The term "Operationally Responsive Space" has an enormous amount of visceral strength to it. "Operational" is full of goodness in and of itself, and "Responsive" sounds almost too good to be true, especially in the space domain where the warfighter's longest running complaint has been, "It won't be there when I need it." In the past couple of years, the ORS crusade has built up a considerable head of steam. The question today is whether ORS is "all it can be." What has happened is that ORS initiatives have generally focused on the acquisition side of responsiveness, and on what industry already knows how to do: build rockets and satellites.

- **ORS is about more than launch.** There is an extraordinary amount of interest within the space community in developing a more responsive launch capability. As discussed earlier, though, it isn't the rocket that determines the responsiveness. In any event, any responsiveness improvements would still provide a capability "too late" for a warfighter in the middle of a firefight.

---

[25] Clark, Stephen. "NASA Retiring TDRS 1 Satellite After More Than 25 Years Of Service," *Spaceflight Now*, October 13, 2009.

- **ORS should concentrate on Tier One.** The current ORS approach looks at responsiveness in terms of "tiers" that are categorized by time to operational status. Tiers Two through Four are focused on an acquisition process ranging from months to decades.

"Alliant Techsystems is preparing to conduct final testing and then will ship the ORS-1 satellite bus, after building it in just 16 months."[26]

What matters to the warfighter, though, is Tier One: what can space do for me right now when I'm in the middle of a firefight and I need help? The warfighter could benefit more readily from improved fusion of data from existing sensors independent of where they are actually located. Tier One looks at how to improve come-as-you-are capabilities by tweaking existing programs and processes. The basic assumption of Tier One is that the most responsive systems are those that are already there. USG investment in ORS capabilities should be directed toward the more near-term opportunities embodied in Tier One.

- **What's missing is the Mission.** One of the reasons ORS is selling itself short is that nobody has defined an actual mission for ORS.

  "The ORS concept was born in 2007 as an effort to field smaller satellites and launch platforms that could quickly be developed and launched to meet urgent national security needs. Officials said several ORS studies are underway as part of the effort to

[26] *SatNews Publishers*, February 18, 2010 http://www.satnews.com/cgi-bin/story.cgi?number=310982287

pinpoint the kinds of space-based capabilities each service would first seek to replace if new combat requirements emerge or existing space platforms become inoperable."[27]

Missions that would be appropriate for a beyond-Teir One ORS capability include quick-reaction engagements (such as show of force or space defense), local surprise (such as unwarned collection or covert operation), and technology demonstrations (such as Space Test Range or space systems experiments). Most importantly, we need to be clear on what it is we are trying to do in the world of more "operational" space.

"Continuing to fund tactical satellites out of budget lines intended directly to serve the tactical war fighter does a disservice to both the taxpayer and the warrior on the ground."[28]
Dr. (LtCol/ret) Mel Tomme,

Measures of Effectiveness (MOE) – Acquisition: During his tenure as Director of the NRO, Keith Hall called for the development of "Seven Standard Yardsticks" to use as comparative metrics for determining the relative value of disparate systems. These Yardsticks could then give a decision-maker a grasp of the relative value of such diverse activities as the value of a covert CIA agent operation compared to the cost of a billion-dollar spy satellite. Comparative metrics that could be useful in such an analysis might include the following MOE.

---

[27] "U.S. Services To Build ORS Costs Into Budget Plans," Defense News, Nov 4, 2009
[28] LtCol Edward B. Tomme, PhD, USAF (Ret.), Research Paper 2006-1, "The Strategic Nature of the Tactical Satellite," Airpower Research Institute, Air University, 2006, p.64.

Does a space program...

- ...provide greater **Utility** than its non-space counterpart? Utility can be a measure of value added, customer demand, or customer commitment.
- ...have sufficient **Political Support** to protect its budget? This support could be at the level of the Combatant Commanders in the Military Service or Agency; in the White House; on the Hill; in the press; or in the eyes of the public – all can influence the success of a program.
- ...have a **Life Cycle Cost** lower than its non-space counterpart? Knowing that Life Cycle Costs are very hard to pin down early in any program, any numbers here must be backed up by substantive evidence to support them.
- ...have an **Annual Budget** that can be protected in yearly budget reviews? Affordability by year in a perennially constrained budget can be a tie-breaker.
- ...have a **Schedule** that is both achievable and includes margin for contingencies?
- ...have a desirable **Technology Readiness Level** (TRL) where you can "buy risk down"?
- ...support and encourage **Net-Centricity**?

Chapter 5

NSS OPERATIONS STRATEGY – PROTECT AND SERVE

INTRODUCTION

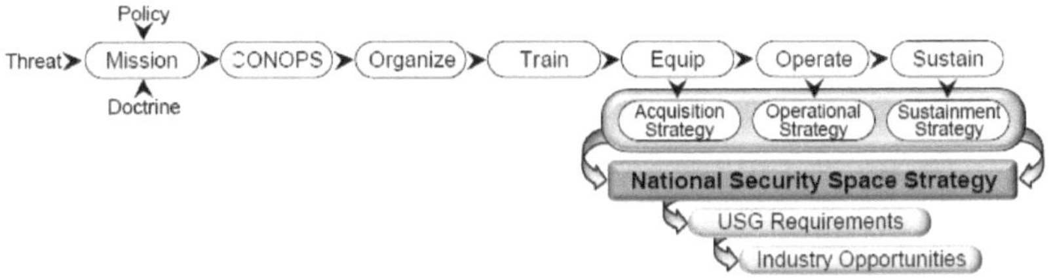

Space Forces provide support other National Security forces cannot. Satellites provide a global network of communications, navigation, meteorology, ISR, and missile warning. Satellites are uniquely able to provide detailed coverage of "denied areas" anyplace on Earth. In a crisis, they are "first on the scene."

Space Forces are constrained in ways other National Security forces are not. Satellites have natural constraints that limit their ability to conduct the space operations listed above.

Most satellite coverage is fleeting. The very first rule to remember is that the basic characteristic of space, "Global Perspective," is not quite as all-encompassing as it sounds. "We're watching you" doesn't usually mean everywhere and all the time (24/7/365). Because of orbital dynamics, most satellites are very limited in how much "global" coverage they can provide. The only satellites that can "hover" over an area of interest are those in geostationary orbits, 22,000 miles above the Earth. Satellites in that orbit are so far from Earth that they can't see much in the way of detail, so they are used primarily for missile warning, communications, large-area weather monitoring, and electronic coverage where the Principle of Instantaneity comes into play. Missile-warning satellites, such as DSP and SBIRS, do provide full-Earth coverage from that altitude using multiple satellites. The improved resolution of SBIRS will enable it to support battlefield characterization in a way that DSP has not. However, electro-optical imaging satellites at that altitude would be much too far away to give the resolution needed by a field commander in a firefight.

Space is not a very efficient "high ground" for tactical reconnaissance. Satellites with "eyes on target" resolution good enough to help the commander can only provide a glimpse of what's around the corner or over the hill, and then, only when (1) over the area of interest, and (2) tasked to watch that particular area.

One of the most pervasive - and misleading - assertions from the user community has long been that "Space won't be there when I need it." Despite its tocsin-like appeal, this charge has only deflected attention from real problems that need to be fixed. The facts are that the user has never enjoyed better support from space and that support is constantly improving. However, it will probably never be all that the user community wants. For many of the services provided from space, military operators and Intelligence Community analysts have long been able to count on instantaneous, omnipresent support. However, in some cases the support the warfighter is claiming to need from space is more readily available from other sources.

There are durability, survivability, and vulnerability issues associated with all satellites. Of course, it is very unlikely that you will hear knowledgeable warfighters complain that they can't get space support when they need it. In fact, the support is so automatic that they are probably not even aware that it is coming from space – which is exactly as it should be. Space support is there today and will be there when the user needs it. Space support is global, ubiquitous, and, in almost all cases, the first force on the scene. Troops will not deploy for combat without coordinating with space forces first.

What needs to be fixed, though, is where space can not provide the support that the users need, in the timeframe that they need it, and with the responsiveness and persistence that they need. This is particularly true in the imagery world of ISR. Which is why the era of the UAV (Unmanned Aerospace Vehicle) [also called UAS (Unmanned Aerospace System), RPV (Remotely Piloted Vehicle), or RPA (Remotely Piloted Aircraft)] is so attractive. For immediate reconnaissance in a firefight, terrestrial forces (UAVs, aircraft and helicopters in the area, even troops on the ground) can usually provide better "high ground" tactical reconnaissance than satellites can.

"U.S. plans to deploy an unmanned surveillance airship to Afghanistan are moving forward, with a contract for the Long Endurance Multi-Intelligence Vehicle (LEMV) demonstration expected to be awarded by year-end. Designed to stay aloft for three weeks carrying a heavy payload of wide-area sensors, the airship is becoming a flagship for Defense Dept. efforts to provide unblinking airborne surveillance to defeat the threat from roadside bombs."[29]

- **Satellites are predictable.** They go fast but don't move well. Space is not very good at "range, speed and maneuverability," the Air Force's mantra for its air fleet. A satellite's "range" is rigidly controlled by its orbit. Its "speed" is immutably tied to laws of orbital physics; "slowing it down" or "speeding it up" will actually cause the reverse effect [you slow a satellite down (take energy out of the orbit) to make it "go faster"; you speed it up (add energy to the orbit) to make it "go slower"]. Any attempt to make a satellite "maneuver" comes at a severe loss of propellant with corresponding loss of orbital lifetime. In essence, any satellite in orbit around the Earth is essentially in a "parking orbit" with all the inherent vulnerabilities that go along with that.

  "Every Trekkie knows you don't do battle in LEO."
  ...anon

- **Satellites are chained to Earth.** Half their brains are on the ground. Operational space forces are characterized by small numbers of very expensive systems that depend upon complex command and control (C2) facilities on Earth. Therefore, any vulnerability calculation for any space system must include the vulnerability of the ground C2 complex, the data processing capability on Earth, and the launch systems that would be used to

[29] *Christopher Fotos, "Pentagon Pushes for Unblinking Surveillance," Aviation Week and Space Technology, Oct 4, 2009.*

replace systems lost to an adversary, as well as the vulnerability of the satellite itself.

- **Satellites are vulnerable.** Their predictable position in space makes them targetable. They can't maneuver much, they can't run, and they can't hide. As National Security Space systems have evolved over the years, our strategy for employing them has evolved as well. We have relied, in the past, upon the assumption that our satellites would not be challenged or attacked in orbit. As our military dependence upon NSS forces has grown, that assumption is becoming less valid. Today, in fact, there are as many as a dozen independent studies underway within DoD on how to win a war without space. The Government has even asked for ideas on how the DoD would fare in a "Day without Space" in which communications and navigation satellites were not available. The issue is: if space isn't going to be there when it's really needed, why spend any money on making it more resistant to attack?

"Some have said that we ought to sharply reduce our dependence on space. That's an overreaction. We do need to look carefully at functions that rely exclusively on any single domain, eliminate single-point failures, and provide alternate paths and sources, always ensuring that architectures are laid out and tested end to end."[30]

[30] Rear Admiral Liz Young, PEO SS, email "Naval Space Acquition Outreach Message, September 17, 2009.

NSS OPERATIONAL STRATEGY: PROTECT & SERVE

The Mission Rules. All NSS operations must start with the Mission. Address operational requirements by looking at the ability of all systems (not just satellites) to accomplish a mission. Satellites can frequently be more useful as a complement to terrestrial capabilities, rather than as their replacement. Look at the services being provided to support a mission, and let the space-air-surface mix be derived from that mission focus. Start with the mission, design for a wartime environment, offload capabilities that can be provided as well from Earth. We can save money and improve effectiveness if we pay for mission durability, not just constellation durability. Know when not to turn to space forces.

MISSION: PROTECT

Mission: Protect our space resources and those of our Allies and friends. Protection requires the ability to 1) deter attacks from taking place, and if deterrence fails, to 2) defend our forces from attack.

Deterrence involves a National commitment. At a minimum, advertise the Nation's willingness to "speak softly but carry a big stick." In today's world of cyber warfare, this could start with a National Policy of Cyber Deterrence, and could include technology demonstrations and operational exercises.

If deterrence fails, be able to respond quickly and decisively. Active protective measures include self-defense, an effective mechanism for battlespace awareness and control, and a credible defensive capability against hostile systems.

- **Build self-defense from within.** In addition to the launch infrastructure - the launch pad, the range tracking facilities as well as the launch vehicle – each satellite program itself consists of three segments: the satellite itself (sometimes called the Space Segment), the communications path from the satellite to wherever its services are being provided (usually called the Communications Segment), and the receiving facility or equipment that is using whatever the satellite provides [called the Receive Segment, the Ground Segment, the User Segment, the Ground Station, the Command and Control (C2) Segment, and several other titles as well, depending upon the program]. Like links in a chain, all four segments must be protected from natural hazards as well as hostile action.

- **Protect the Launch Segment where possible.** Protecting the launch infrastructure is an inordinately complex issue. Turbulent weather and other natural phenomena are daunting in themselves – hostile threats only compound the challenge. Proliferation, redundancy, and backup capabilities all contribute to minimize dependence upon any one facility or rocket. To the extent that cooperative programs can be pursued with international launch agencies, the development of compatible launch, range and support capabilities enhance the resiliency and durability of our launch capability.

- **Harden and proliferate the Ground Segment.** Protecting facilities on Earth is relatively straightforward. We have had a lot of experience building secure, earthquake-resistant facilities, hardened receive terminals, and the like.

- **Distribute the Communications Segment.** The Communications Segment can include telephone lines, satellite links, microwave towers, fiber optic links, and all other communications paths that are out there today. Here again, it is only a matter of determination and money to provide alternate communication paths and hardened communication links.

- **Harden and diversify the Space Segment.** As described in the Threat section of this monograph, satellites are vulnerable to any number of natural and hostile effects. Nevertheless, self-defense options are available. The issue, then, becomes how to protect satellites from natural hazards as well as attack, both kinetic and non-kinetic. For all sorts of very practical, logical, technical reasons, it just isn't possible to use a kinetic ASAT to chase down another kinetic ASAT from behind. Neither is it very realistic to rely, in space, on the terrestrial force defensive concept known as "trading space for time." Most non-kinetic attacks can be prosecuted at the speed of light, which doesn't give a satellite any time to "retreat" as the attack is instantaneous. The solution, of course, is that in order to be available when needed, defensive systems must be internal to on-orbit spacecraft. Rather than attempting to "trade space for time," a more effective approach would be the space equivalent of "defense in depth," where the satellite employs distributed payloads; alternate communication paths; internal, built-in, automatic defenses; and the "assured response" of "equivalent" rather than "in kind" retaliation. Decoys, movement to avoid an incoming attack, shields to protect against laser blinding, and other defensive mechanisms could be in every satellite's protective arsenal – at a price.

- **Provide effective battlespace awareness and control.** In order to provide an effective defense, space forces must be able to understand and monitor all activities in space leading to a comprehensive understanding of what is there, as well as what each satellite intends to accomplish.
- **You can't kill the enemy you can't see.** The #1 priority for Air Force Space Command has long been an improved ability to see what is going on in space. Improved Space Situational Awareness (SSA) is vital in order to understand the operational implications of unexplained movements of enemy satellites, potential collisions with space debris, and other basic Space Order of Battle characteristics. Counter-space systems must have effective detection, identification, tracking and Battle Damage Assessment (BDA) capabilities in order to accomplish their mission.
- **Plan for an enemy to attack NSS <u>systems</u> and not just satellites.** There is no obvious, compelling requirement that offensive weapons need to be based <u>in</u> space in order to kill satellites. Because of the extended tether that ties every satellite to an extremely valuable, vulnerable facility on the ground, most satellites can be disabled by targeting the most vulnerable element of the C2 chain and not necessarily the satellite itself. Kinetic and cyber attacks can be prosecuted on the ground, for example, where the enemy's mission control station, C2 links, or data processing facilities can be disrupted or destroyed. In scenarios where the satellite is in fact the weakest link, counter-space systems must be able to achieve non-kinetic "kills" that won't add debris to the space

environment. In either case, be wary of space-weapon advocates who overlook the system-of-systems implications.

- **Prepare the Operator <u>and</u> the User to expect an Asymmetrical Attack.** Waging war in space is neither cheap nor easy. The cost of getting to space and sustaining operations there is so high that access to space is essentially prohibitive to most aggressors. That does <u>not</u> mean, though, that satellites are less vulnerable to disruption and damage than terrestrial forces. In fact, there are many ways for an adversary to disrupt space operations without having to be in space themselves. Asymmetrical military operations are becoming increasingly common on Earth. Somali pirates use small watercraft to capture giant oil tankers. The Iraqis used Toyota pickups in their invasion of Kuwait. Almost any radical can figure out how to fly small UAVs that can carry disproportionate WMDs. Space operations are likely to become an increasingly desirable target for adversaries who can apply asymmetrical force to achieve their goals.
- **Anticipate cyber interference as the first level of attack.** It is usually non-attributable, at least in the first stages of a campaign, and the penalty for failure is relatively minimal and plausibly denied.
- **Anticipate terrestrial interference as the second level of attack.** Operational planners must assume an adversary will try to interfere with a vital terrestrial node. It is relatively low-technology and can be disguised as "accidental" or "unintentional."

MISSION: SERVE

Mission: Provide "Beat the Need" responsiveness to all users of space services. Integrate cross-domain resources to serve space users with responsiveness never before seen.

Responsiveness Starts on the Ground. Given the characteristics of forces in orbit, more responsive NSS Operations begins with a robust ground system that supports the satellites and delivers mission data to the users quickly and efficiently.

Exploit "Big Space" for its untapped responsiveness. "Big Space" is an essential part of NSS forces. Big Space has provided the bulk of the euphemistic "National Technical Means (NTM) of Verification" and strategic systems such as GPS, Milsatcom, weather monitoring, and missile warning (DSP and SBIRS). These systems and their predecessors helped the U.S. win the Cold War. Big Space will continue to play a key role in strategic Indications and Warning (I&W), strategic monitoring, and other surveillance roles where access is denied to forces on Earth or where time-sensitive global reach is required. While Big Space has traditionally been limited in its persistence, immediacy, durability, survivability, and responsiveness, its capabilities have really not been fully exploited. This is a political obstacle rather than a technical engineering challenge. The needs of today's users - warfighters and others alike - demand that the floodgates for Big Space processes and products be opened wide. Big Space can contribute much more than it is being tasked to provide today.

Use terrestrial forces for tactical support. In the tactical support world, Big Space should be relied upon only for that which Earth-based forces cannot provide. Big Space has largely been marginalized for tactical support – especially imagery. Therefore, do not fund Big Space improvements advertised as tailored to the operational warfighter for tactical support. Space forces, "Big Space" as well as "Little Space," should be designed for, and limited to, the tasks for which they are best suited. It is wasteful to spend money on space programs that provide more than is required or on space activities that could be done more effectively by other means. In particular, stop funding ISR programs that advertise increased persistence and responsiveness from space. Persistence must be built into operational forces by fusing data from the full spectrum of ISR sources available. Earth-based options are usually more responsive, persistent, and cost-effective than satellites are for intra-theater type support

Measures of Effectiveness (MOE) – Operational.

Are the Users...

- ...getting the support they need when and where they need it?
- ..."in charge" and able re-task space assets to accommodate a fluid engagement?
- ...getting space support without having to worry about it any more than they worry about support they are getting from other forces?

Chapter 6

NSS SUSTAINMENT STRATEGY – STRENGTHEN THE FUTURE

INTRODUCTION

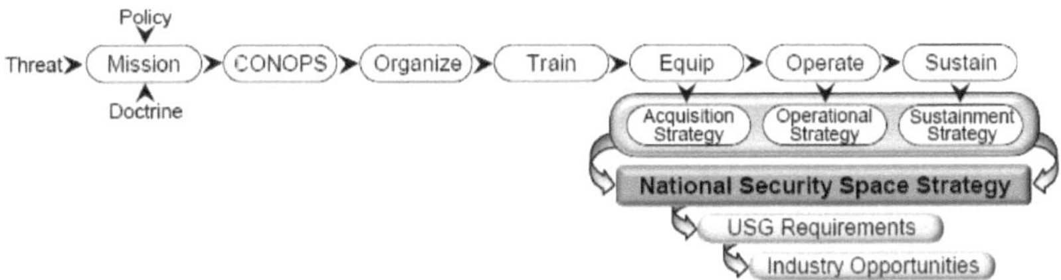

Space Systems Don't Die Hard. For at least four decades, the space community has marveled at the continuing phenomenon of long-lived satellites. Most of the DoD and Intelligence Community constellations have satellites that last two to three times their expected lifetime. These space systems are surviving into their AARP years as they gracefully degrade and lose individual components and subsystems along the way. As a result, almost all constellations include "legacy" satellites that are limping along but still perform some of their missions and can help in a crisis. This phenomenon is a result of the "bathtub curve," documented in

NASA's Space Systems Handbook, and occurs once the space system has survived early threats such as launch and initial environmental stresses. Once past the front end of their "bathtub curve," satellites tend to need less care and feeding for most of their lives. Eventually, their maintenance requirements gradually increase. Many of them remain at least partially productive until they finally give up the ghost.

"Look-ahead" Sustainment is a Force Multiplier. The longest-lasting element of an integrated NSS strategy, Sustainment keeps a space force viable once it has been designed, built, launched, and put into operation. Sustainment applies to both the USG side as well as the industry side. How to keep up with a changing threat, how to maintain the systems on the ground, how to find new and better ways for satellites to accomplish their mission, how to adapt as technology improves, etc., are all aspects of a robust sustainment strategy.

SUSTAINMENT STRATEGY: STRENGTHEN THE FUTURE

It is to our advantage to prolong the lives of our satellites. We have grown critically dependent upon the services we get from space. New forces take longer to come into the inventory than we want or need. Delay is inevitable under our current processes. Until our acquisition system changes, we aren't going to get replacement systems any faster than we do today. What we have to do, then, is find ways to keep existing programs healthy while we develop their replacements.

Under these "facts of life," a responsible <u>strategy</u> for NSS Sustainment includes a look-ahead concept:

Maximize residual value of existing constellations while paving the way for an efficient transition to next-generation systems.

This requires a "look-ahead" approach that recognizes inescapable aspects of today's space forces:

- Satellites already on orbit can usually contribute to their mission long after they reach their "designed" end of life. Protect today's operations. Ensure that current space capabilities do not degrade.
- Satellites intended to replace those already on orbit tend to take longer to deploy than originally budgeted or forecast. It is possible to have potentially dangerous gaps in service if the satellites already on orbit can't cover the mission until the new satellites arrive. Expect the arrival of new systems to be later than budgeted or forecast. Launch success is never a guarantee. Protect transition between acquisition and sustainment.
- Next-generation satellites tend to bring orders-of-magnitude improvement in capability, hence the desirability of getting them up and operating as soon as possible. As part of that process, shorten the development timeline by getting closer to the user and looking for ways to apply valuable operational feedback to the development of next-generation systems.
- The delays in getting next-generation satellites operational are usually so great that the requirements that led to their performance parameters in the first place have largely been overtaken by events, such that, unless otherwise accounted for, it is possible for satellites to be "obsolete" on the day of launch.

Three Steps to a long-term program for sustaining America's Space Force:

1. **In Acquisition, build sustainability in up front**. Major design objectives consistent with this Sustainment strategy include flexible programming, payload flexibility, and modular design.
2. **In Operations, reward programs that protect the mission while reducing O&M costs.** The USG is almost always justified in requiring an incumbent to project a declining cost for continuing to operate a long-term program. Learning-curve improvements, value engineering savings, and continuing management efficiencies are all potential sources of reducing O&M costs. This is a win-win situation: the USG saves money over time, and the supplier provides a visible commitment to the success of their program over the long haul, which in turn makes them more desirable for follow-on procurements.
3. **In Sustainment, protect people and facilities.** Of course, Sustainment carries with it a commitment to longer-term issues than just cost. Sustainment considerations also include personnel and facilities.

Good people are <u>not</u> hard to find. Just as with other military and intelligence elements, space forces need a full spectrum of experts. Experienced operators, seasoned acquisition experts, radical technologists, software geniuses, consummate industrial professionals, and many other specialties are all in demand. There is no end in sight to the requirements for space experts. There have been many speeches and articles in the aerospace trade journals about how the space industry can't get the people we need to do the job. What has happened is that space has gradually evolved from a high-tech leading-edge thrill-ride that attracted the best-qualified people and received enough funding to tackle impossible technical challenges. As the space industry has become more mainstream, and as the products have become more commodity-like, space is no longer the employer of choice. It now has to compete for dollars and good people. Many in the industry are simply not prepared to be able to do that while others are adapting quickly. Those who have been able to adjust have been successful in finding the people and dollars they need for their programs. SpaceX, for example, represents the thrill-ride side – daring to press the boundaries of what traditional manufacturers said "couldn't be done." This is clearly attracting interest. It is noteworthy that in Q&A after SpaceX presentations in symposia across the country, the most often-asked question is, "Are you hiring?" This is even asked by rocket scientists who are already employed. As part of the conversion to United Launch Alliance from separate EELV facilities, Boeing had an immediate requirement to move their production line from Seal Beach, California to Waterton, Colorado. By using innovative incentives and tapping into new university graduates, they met their hiring goals within six months. The fact is that when the need is real, and the contracts are signed, we are doing much better than many predicted. Those who actually have the problem have found solutions that work, using the age-old laws of supply and demand.

Space forces cannot afford to have facilities fall behind. Facilities for space R&D, testing, and manufacture are generally either owned and operated by industry, or owned by the Government and operated by industry. In either case, it is to our collective advantage to transform facilities as technology allows. There are examples around the world where facilities have successfully leapfrogged into next-generation designs, even during recessions, with demonstrable payback.

Measures of Effectiveness (MOE) – Sustainment.

Is the program...

- ...reducing O&M cost without losing mission capability?
- ...encouraging a strong Technology Base?
- ...workforce motivated, with a visible career progression plan?
- ...eco-friendly with a funded program to reduce the eco-footprint?
- ...self-sustaining, i.e., no bail-outs required?
- ...providing international leadership and setting standards of performance that are unreachable by foreign competitors?
- ...infrastructure responsive to changing requirements – anticipatory where possible, quick-reacting when needed – with built-in mechanisms for forecasting future sustainability requirements based upon current acquisition and operational environments?
- ...looking ahead to threats, requirements, trends and technologies beyond the current contract?

Chapter 7

NEEDS AND OPPORTUNITIES – A PARTNERSHIP FOR PROGRESS

INTRODUCTION

Strategy is only as effective as its implementation. This monograph has developed a three-pronged approach for National Security Space Strategy that encompasses Acquisition, Operations and Sustainment. Implementing this approach requires two more steps. First is identifying what the U.S. Government "needs" to do in order to implement National Security Space Strategy. Second is deriving Industry "opportunities" from those "needs." As you will see below, balancing the two can create an effective partnership for progress.

U.S. GOVERNMENT NEEDS

Implementing NSS Strategy won't be done overnight. The key to effective implementation is recognizing that strategy is an extended, continuing process that is constantly gathering new data, adjusting to changing conditions, shedding old concepts, and moving forward with new ideas. What Strategy must focus on is staying relevant as operational demands change across the National Security domain.

"Needs" are not the same as "Requirements." Even though it may seem intuitively obvious that the Government "needs" something, there is no guarantee that need will ever be certified as a formal "requirement" and funded in the President's Budget. Even when requirements are formalized in the DoD's budget process, "tinkering" with the requirement usually causes overruns, late schedules, and considerable angst on everyone's part. It is unrealistic in today's process to expect the government to be able to reduce the turbulence in requirements.

We need to find better ways to define needs and requirements. What might help would be a greater effort made to mesh the technology of "what can be accomplished" with "what is needed." This knowledge would be critical to helping the USG defer what is not doable and keep it in the high-risk category. NSS Strategy must encourage operational users to consolidate requirements and then run them by technology leaders to ensure that the needs can be met within a reasonable time and financial framework.

The Strategy of Protect and Serve has clear needs associated with it. Many of these needs have already been validated and funded.

- **Improve our ability to know what else is going on up there.** Without a better Space Situational Awareness capability, we might as well be whistling in the dark. Data from Space Fence and other sensors, on the ground and in space, needs to be fused into a

Common Operational Picture (COP) that takes the guesswork out of space operations.

- **Improve battlespace awareness and control.** An improved Joint Space Operations Center that mirrors other Air Operations Centers and provides high-quality decision-making information is essential. The JMS Program [Joint Space Operations Center (JSpOC), Mission, System] must be implemented without delay.
- **Improve Mission Assurance by buying one EELV ahead.** A single EELV "in the barn" will give today's DoD and Intelligence Community space programs all the backup they need for their constellations.
- **Improve UAV effectiveness.** Show the National Security apparatus how to use space to exploit the dramatic evolution of the UAV as an operational warfighting asset. Data fusion from space and Unmanned Air Vehicles could revolutionize operational support.
- **Protect our space assets by building "GEO-Cops."** In the event that the space environment becomes too hostile to support friendly forces on Earth, it may be desirable to deploy laser or other speed-of-light weapons in space and task them with the job of protecting the satellites they can see. In order to gain the maximum visibility for these "armed escorts," it would be desirable to station them in geosynchronous earth orbit (GEO), hence the name "GEO-Cops."
- **Expand the envelope.** We can do <u>a lot</u> better with what we have. However, until we quit believing that bigger is the only way to go to get better and that new systems are the only way to improve the effectiveness

of National Security Space operations, we are stuck in an expensive circle. NGA's growing acceptance of commercial imagery, and the launch industry's expansion with the addition of SpaceX are examples where stretching the envelope can produce new ideas and new approaches. DARPA's "fractionation" effort, an attempt to fly subsystems in formation to act like a single satellite, is another.[31] We need to do more in other areas, particularly in cyber defense, data fusion, TSAT-like internet-in-the-sky breakthroughs, and exploitation of the Principle of Instantaneity.

INDUSTRY OPPORTUNITIES

Industry opportunities abound and are as challenging as ever. The American space industry has conquered virtually every challenge it has ever been given. Satellites today are making huge contributions to National Security. The next challenge is to contribute to missions in the tactical arena. Two requirements – timeliness and responsiveness – remain beyond the limits of today's space systems. Their solution is likely to involve data sharing and integration at levels never before seen.

Other industry opportunities are limited only by our imagination. Contrary to popular belief, funding and support will always be available for a good idea. The key, of course, is proving the "good."

---

[31] "Breaking up may be good to do," The Space Review, November 2, 2009

Chapter 8

NINE RED HERRINGS AND THE BOTTOM LINE

INTRODUCTION

The "**Bottom Line**" of this paper is in two parts. First, is a summary of what we call "**Nine Red Herrings**" that have been inhibiting the development of an improved, sound, NSS Strategy. By understanding the logic flaws in these assertions, it becomes easier to understand the second part, which summarizes the **National Security Space Strategy Considerations** presented in this monograph.

NINE RED HERRINGS

For over 50 years, the U.S. has sustained a strong aerospace force for the Nation and a strong aerospace market for industry. The difficulty on the part of national authorities to advocate a clear strategy for National Security Space comes about, at least in part, because others are pushing concepts that simply are not true. As discussed in this monograph, these "**Nine Red Herrings**" can be summarized as follows:

1. **Red Herring:** "The U.S. does not have a National Space Strategy."

TRUTH: Sure we do! It may not have been articulated well or enunciated formally, but we have been following it for the past 50 years. What it needs now is a makeover – which is the purpose of this monograph.

2. **Red Herring:** "ORS will make space more operationally responsive."

TRUTH: Not the way we're going. We are not targeting ORS at the operational level. We need to put real money into Tier One. What makes that difficult for the space community to accept is that much of Tier One involves changes in existing processes and is likely to require non-hardware, and even non-space, solutions.

3. **Red Herring:** "We need weapons in space to protect our satellites."

TRUTH: Not really, and even if we did have them, it is unlikely they would be able to respond in time to do anything besides clutter up orbits. We need to be able to protect our satellites. However, that protection need not be in space. Again, this solution won't be particularly appealing to the build-things-in-space community.

4. **Red Herring:** "Space is not there for us."

TRUTH: That complaint may have been true in the past. However, today space support is ubiquitous around the globe [24 / 7 / 365]. Space is first on the scene of every new crisis. What still needs to be fixed is improving timeliness and responsiveness by linking space with terrestrial complements...yet another non-space solution.

5. **Red Herring:** "The space acquisition process is broken."

TRUTH: Only in the sense of our reluctance (and occasionally our overt resistance) to put the necessary changes in place to fix problems that are in fact fixable.

6. **Red Herring:** "One-of-a-kind ('Big Space') platforms are unworkable."

TRUTH: Only for the missions that Big Space shouldn't be considered for in the first place.

7. **Red Herring:** "Launch isn't responsive enough for the warfighter."

TRUTH: Launch is responsive enough for all the support that it is capable of providing. We fall into a trap of our own making when we try to include launch as a solution to "hours or days" responsiveness. Strategic ICBM and SLBM forces are the only launch systems that can meet "minutes or hours" responsiveness requirements. Besides being virtually impossible, "launch on demand" is unrealistic and prohibitively expensive.

8. **Red Herring:** "Space must be a separate military Service."

TRUTH: It may help but it certainly doesn't have to be. Reorganization can break as many (or more) things than it fixes.

9. **Red Herring:** "The aerospace industry can't get the people we need to do the job."

TRUTH: We know that's not true. It is all a matter of supply and demand. Those who have stepped up to the challenge have been, and will continue, to be successful in getting the right people for the right jobs.

National Security Space Strategy – THE BOTTOM LINE

The Nation's strategy for National Security Space (NSS) programs started fifty years ago with President Eisenhower's two basic principles: "Freedom of Space," and "Space for Peaceful Purposes." NSS programs have pushed the envelope of "do-ability" ever since then, leading to today's architecture of extremely capable contributions to the Nation's security. For 50 years, NSS strategy has taken advantage of the "high ground" of space by building and deploying satellites - using the best technological solutions possible - and getting every last ounce of performance out of them.

Today, U.S. and international space programs are absolutely vital to every military operation, every commercial undertaking, and every financial transaction worldwide. National Security Strategy, however, has not kept up with the increased demand, the burgeoning dependence, and the growing threats to space operations. In response, this monograph recommends improved NSS Strategy based upon the following "OV-1" overview. As shown, a "necessary and sufficient" Strategy includes separate but integrated strategies for acquisition, operations and sustainment:

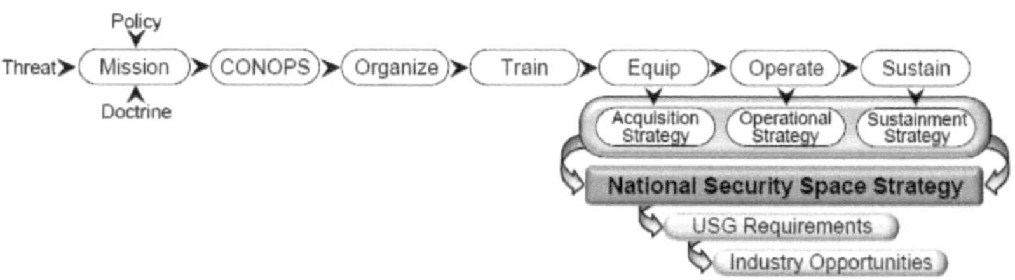

NSS Acquisition Strategy - Lead Better, Follow Well, Buy Smart. Acquisition of space systems is an extraordinarily difficult task. This monograph presents several "lessons learned" in how to "buy smart":

- The USG should prioritize program decisions based upon criticality.
- USG acquisition centers should go "Back to Basics" – and stay there.
- The USG should work toward a <u>balance</u> in acquisition.
- The USG should design for the <u>long haul</u>.
- The USG should <u>intensify</u> ORS by focusing on Tier One.

NSS Operational Strategy – Protect and Serve. The growing importance of space operations has reemphasized a responsibility for Space Forces that is similar to that of a neighborhood community police department – Protect and Serve – as embodied in two missions:

- **Mission**: Protect our space resources and those of our Allies and friends.
- **Mission**: Provide "Beat the Need" responsiveness to all users of space services.

NSS Sustainment Strategy – Strengthen the Future. Government and industry need to work together to ensure a strong space force despite growing threats and continuing fiscal uncertainty. The bottom line is that space professionals must implement three steps that lead to "Look-Ahead Sustainment" and build a long-term program for sustaining America's Space Force:

- In **Acquisition**, build in sustainability up front.
- In **Operations**, reward programs that protect the mission while reducing O&M costs.
- In **Sustainment**, protect people and facilities. Good people are not hard to find, and space forces cannot afford to have facilities fall behind.

National Security Space Strategy:
Protect and Serve

APPENDIX A
BIBLIOGRAPHY

"America's Leadership in Space," letter to President Obama from the Committee for U.S. Space Leadership, March 10, 2009

AU-18 Space Primer, ACSC, Air University Press, Maxwell AFB, Alabama, 9/09. pg 39.

"Breaking up may be good to do," The Space Review, November 2, 2009

Clark, Stephen. "NASA Retiring TDRS 1 Satellite After More Than 25 Years Of Service," *Spaceflight Now*, October 13, 2009.

"Chief of staff highlights importance of space to Air Force mission," Defense Media Activity-San Antonio, February 19, 2010.

Commercial Space Launch Policy, Sep 1990

Department of Defense Space Policy, July 09, 1999.

Fotos, Christopher, "Pentagon Pushes for Unblinking Surveillance," Aviation Week and Space Technology, Oct 4, 2009.

Garretson, Peter, "Elements of a 21st century space policy," The Space Review. Aug 3, 2009.

Global Space Activity Revenues and Budgets, 2007. The Space Report. www.thespacereport.org

Grant, Rebecca. "Insecurity in Space," *Air Force Magazine*, October, 2009

High Frontier, US Air Force Space Command Journal.

Hill. Jeffrey, "GAO Report Highlights Issues with U.S. Commercial Launch Policy," Satellite TODAY, published by Access Intelligence, LLC, 4 Dec 2009

http://www.satellitetoday.com/st/headlines/32978.html

Making Strategy: An Introduction to National Security Processes and Problems, Chapter 11, Published 1988 by Air University Press. August 1988.

McDonald, Robert A., ed., Corona Between the Sun and the Earth, The First NRO Reconnaissance Eye in Space (Bethesda, MD: American Society for Photogrammetry and Remote Sensing, 1997).

"National Security Space Launch Report," The Congressionally Mandated National Security Space Launch Requirements Panel, RAND National Defense Research Institute, 2006.

National Space Transportation Policy, Aug 1994

Paone, Chuck. "Leaders call for balance between network use and protection," 66th Air Base Wing Public Affairs, Air Force Materiel Command, reporting on comments by Air Force Electronic Systems Center Commander Lt. Gen. Ted Bowlds at the 7th annual Net-centric Operations Conference, 24 Sep 2009 in New Castle NH.

"Space: The Highest Ground," Strategic Forecasting, Inc. (STRATFOR), October 19, 2009

Satellite News, December 4, 2009

SatNews Publishers, February 18, 2010

Taverney, Thomas D. and James D. Rendleman. Ten Rules for Common Sense Space Acquisition. High Frontier, USAF Space Command, Vol 6. # 1.

Tirpak, John A., "The Doctor's Prescription," Air Force Magazine.com, September 14, 2009

Tomme, LtCol Edward B. PhD, USAF (Ret.), Research Paper 2006-1, "The Strategic Nature of the Tactical Satellite," Airpower Research Institute, Air University, 2006.

Tucker, Christopher K. "The Watchmen and the Scientists," Science Progress, Nov 20, 2009.

Young, A. Thomas , Chairman, "Leadership, Management, and Organization for National Security Space (NSS) – Report to Congress of the Independent Assessment Panel on the Organization and Management of National Security Space," Institute for Defense Analyses, IDA Group Report GR-69, July, 2008.

Young, Liz PEO Rear Admiral SS, email "Naval Space Acquition Outreach Message, September 17, 2009.--U.S. National Space Policy, August 31, 2006

"U.S. Services To Build ORS Costs Into Budget Plans," Defense News, Nov 4, 2009

"Washington Watch," Air Force Magazine, November, 2009

www.heavens-above.com

www.weather.com

http://en.wikipedia.org/wiki/Principles_of_War#United_States_principles_of_war

APPENDIX B
UNITED STATES PRINCIPLES OF WAR[32]

(Refer to U.S. Army Field Manual FM 3-0)

The United States Armed Forces use the following nine principles of war in training their officers:

Objective - Direct every military operation toward a clearly defined, decisive and attainable objective. The ultimate military purpose of war is the destruction of the enemy's ability to fight and will to fight.

Offensive - Seize, retain, and exploit the initiative. Offensive action is the most effective and decisive way to attain a clearly defined common objective. Offensive operations are the means by which a military force seizes and holds the initiative while maintaining freedom of action and achieving decisive results. This is fundamentally true across all levels of war.

Mass - Mass the effects of overwhelming combat power at the decisive place and time. Synchronizing all the elements of combat power where they will have decisive effect on an enemy force in a short period of time is to achieve mass. Massing effects, rather than concentrating forces, can enable numerically inferior forces to achieve decisive results, while limiting exposure to enemy fire.

Economy of Force - Employ all combat power available in the most effective way possible, allocate minimum essential combat power to secondary efforts. Economy of force is the judicious employment and distribution of forces. No part of the force should ever be left without purpose. The allocation of available combat power to such tasks as limited

[32] http://en.wikipedia.org/wiki/Principles_of_War#United_States_principles_of_war

attacks, defense, delays, deception, or even retrograde operations is measured in order to achieve mass elsewhere at the decisive point and time on the battlefield.

Maneuver - Place the enemy in a position of disadvantage through the flexible application of combat power. Maneuver is the movement of forces in relation to the enemy to gain positional advantage. Effective maneuver keeps the enemy off balance and protects the force. It is used to exploit successes, to preserve freedom of action, and to reduce vulnerability. Continually it poses new problems for the enemy by rendering his actions ineffective, eventually leading to defeat.

Unity of Command - For every objective, seek unity of command and unity of effort. At all levels of war, employment of military forces in a manner that masses combat power toward a common objective requires unity of command and unity of effort. Unity of command means that all the forces responsible are under one command. It requires a single command with the requisite authority to direct all forces in pursuit of a unified purpose.

Security - Never permit the enemy to acquire unexpected advantage. Security enhances freedom of action by reducing vulnerability to hostile acts, influence, or surprise. Security results from the measures taken by a command to protect his forces. Knowledge and understanding of enemy strategy, tactics, doctrine, and staff planning Improve the detailed planning of adequate security measures.

Surprise - Strike the enemy at a time or place or in a manner for which he is unprepared. Surprise can decisively shift the balance of combat power. By seeking surprise, forces can achieve success well out of proportion to the effort expended. Surprise can be in tempo, size of force, direction or location of main effort, and timing. Deception can aid the probability of achieving surprise.

Simplicity - Prepare clear, uncomplicated plans and concise orders To ensure thorough understanding. Everything in war is very simple, but the simple thing is difficult. To the uninitiated, military operations are not difficult. Simplicity contributes to successful operations. Simple plans and clear, concise orders minimize misunderstanding and confusion. Other factors being equal, parsimony is to be preferred.

www.ingramcontent.com/pod-product-compliance
Lightning Source LLC
Chambersburg PA
CBHW031240280526
45784CB00004B/1655

9780557317745